T0235800

Federated Learning

Synthesis Lectures on Artificial Intelligence and Machine Learning

Editors
Ronald J. Brachman, *Jacobs Technion-Cornell Institute at Cornell Tech*
Francesca Rossi, *IBM Research AI*
Peter Stone, *University of Texas at Austin*

Federated Learning

Qiang Yang, Yang Liu, Yong Cheng, Yan Kang, Tianjian Chen, and Han Yu

ISBN: 978-3-031-00457-5   paperback
ISBN: 978-3-031-01585-4   ebook
ISBN: 978-3-031-00030-0   hardcover

DOI 10.1007/978-3-031-01585-4

A Publication in the Springer series
SYNTHESIS LECTURES ON ARTIFICIAL INTELLIGENCE AND MACHINE LEARNING

Lecture #43
Series Editors: Ronald J. Brachman, *Jacobs Technion–Cornell Institute at Cornell Tech*
                Francesca Rossi, *IBM Research AI*
                Peter Stone, *University of Texas at Austin*
Series ISSN
Synthesis Lectures on Artificial Intelligence and Machine Learning
Print 1939-4608   Electronic 1939-4616

Federated Learning

Qiang Yang
WeBank and Hong Kong University of Science and Technology, China

Yang Liu
WeBank, China

Yong Cheng
WeBank, China

Yan Kang
WeBank, China

Tianjian Chen
WeBank, China

Han Yu
Nanyang Technological University, Singapore

SYNTHESIS LECTURES ON ARTIFICIAL INTELLIGENCE AND MACHINE LEARNING #43

ABSTRACT

How is it possible to allow multiple data owners to collaboratively train and use a shared prediction model while keeping all the local training data private? Traditional machine learning approaches need to combine all data at one location, typically a data center, which may very well violate the laws on user privacy and data confidentiality. Today, many parts of the world demand that technology companies treat user data carefully according to user-privacy laws. The European Union's General Data Protection Regulation (GDPR) is a prime example. In this book, we describe how federated machine learning addresses this problem with novel solutions combining distributed machine learning, cryptography and security, and incentive mechanism design based on economic principles and game theory. We explain different types of privacy-preserving machine learning solutions and their technological backgrounds, and highlight some representative practical use cases. We show how federated learning can become the foundation of next-generation machine learning that caters to technological and societal needs for responsible AI development and application.

KEYWORDS

federated learning, secure multi-party computation, privacy preserving machine learning, machine learning algorithms, transfer learning, artificial intelligence, data confidentiality, GDPR, privacy regulations

Contents

Preface

This book is about how to build and use machine learning (ML) models in artificial intelligence (AI) applications when the data are scattered across different sites, owned by different individuals or organizations, and there is no easy solution to bring the data together. Nowadays, we often hear that we are in the era of big data, and big data is an important ingredient that fuels AI advances in today's society. However, the truth is that we are in an era of small, isolated, and fragmented data silos. Data are collected and located at edge devices such as mobile phones. Organizations such as hospitals often have limited views on users' data due to their specialties. However, privacy and security requirements make it increasingly infeasible to merge the data at different organizations in a simple way. In such a context, federated machine learning (or federated learning, in short) emerges as a functional solution that can help build high-performance models shared among multiple parties while still complying with requirements for user privacy and data confidentiality.

Besides privacy and security concerns, another strong motivation for federated learning is to maximally use the computing power at the edge devices of a cloud system, where the communication is most efficient when only the computed results, rather than raw data, are transmitted between devices and servers. For example, autonomous cars can handle most computation locally and exchange the required results with the cloud at intervals. Satellites can finish most of the computation for information that they are to gather and communicate with the earth-based computers using minimal communication channels. Federated learning allows synchronization of computation between multiple devices and computing servers by exchanging only computed results.

We can explain federated learning with an analogy. That is, an ML model is like a sheep and the data is the grass. A traditional way to rear sheep is by buying the grass and transferring it to where the sheep is located, much like when we buy the datasets and move them to a central server. However, privacy concerns and regulations prevent us from physically moving the data. In our analogy, the grass can no longer travel outside its local area. Instead, federated learning employs a dual methodology. We can let the sheep graze multiple grasslands, much like our ML model that is built in a distributed manner without the data traveling outside its local area. In the end, the ML model grows from everyone's data, just like the sheep feed on everyone's grass.

Today, our modern society demands more responsible use of AI, and user privacy and data confidentiality are important properties of AI systems. In this direction, federated learning is already making significant positive impact, ranging from securely updating user models on mobile phones to improving medical imaging performance with multiple hospitals. Many existing works in different computer science areas have laid the foundation for the technology, such as distributed optimization and learning, homomorphic encryption, differential privacy, and secure multi-party computation.

There are two types of federated learning, horizontal and vertical. The Google GBoard system adopts horizontal federated learning and shows an example of B2C (business-to-consumer) applications. It can also be used to support edge computing, where the devices at the edge of a cloud system can handle many of the computing tasks and thus reduce the need to communicate via raw data with the central servers. Vertical federated learning, proposed and advanced by WeBank, represents the B2B (business-to-business) model, where multiple organizations join an alliance in building and using a shared ML model. The model is built while ensuring that no local data leaves any sites and maintaining the model performance according to business requirements. In this book, we cover both the B2C and B2B models.

To develop a federated learning system, multiple disciplines are needed, including ML algorithms, distributed machine learning (DML), cryptography and security, privacy-preserving data mining, game theory and economic principles, incentive mechanism design, laws and regulatory requirements, etc. It is a daunting task for someone to be well-versed in so many diverse disciplines, and the only sources for studying this field are currently scattered across many research papers and blogs. Therefore, there is a strong need for a comprehensive introduction to this subject in a single text, which this book offers.

This book is an introduction to federated learning and can serve as one's first entrance into this subject area. It is written for students in computer science, AI, and ML, as well as for big data and AI application developers. Students at senior undergraduate or graduate levels, faculty members, and researchers at universities and research institutions can find the book useful. Lawmakers, policy regulators, and government service departments can also consider it as a reference book on legal matters involving big data and AI. In classrooms, it can serve as a textbook for a graduate seminar course or as a reference book on federated learning literature.

The idea of this book came about in our development of a federated learning platform at WeBank known as Federated AI Technology Enabler (FATE), which became the world's first open-source federated learning platform and is now part of the Linux Foundation. WeBank is a digital bank that serves hundreds of millions of people in China. This digital bank has a business alliance across diverse backgrounds, including banking, insurance, Internet, and retail and supply-chain companies, just to name a few. We observe firsthand that data cannot be easily shared, but the need to collaborate to build new businesses supported by ML is very strong.

Federated learning was practiced by Google at large-scale in its mobile services for consumers as an example of B2C applications. We took one step further in expanding it to enable partnerships between multiple businesses in a partnership for B2B applications. The horizontal, vertical, and transfer learning-based federated learning categorization was first summarized in our survey paper published in *ACM Transactions on Intelligent Systems and Technology (ACM TIST)* [Yang et al., 2019] and was also presented at the 2019 AAAI Conference on Artificial Intelligence (organized by the Association for the Advancement of Artificial Intelligence) in Hawaii. Subsequently, various tutorials were given at conferences such as the 14th Chinese Computer Federation Technology Frontier in 2019. In the process of developing this book, our open-source federated learning system, FATE, was born and publicized [WeBank FATE, 2019] (see https://www.fedai.org), and the first international standard on federated learning via IEEE is being developed [IEEE P3652.1, 2019]. The tutorial notes and related research papers served as the basis for this book.

Qiang Yang, Yang Liu, Yong Cheng, Yan Kang, Tianjian Chen, and Han Yu
November 2019, Shenzhen, China

Acknowledgments

The writing of this book involved huge efforts from a group of very dedicated contributors. Besides the authors, different chapters were contributed by Ph.D. students, researchers, and research partners at various stages. We express our heartfelt gratitude to the following people who have made contributions toward the writing and editing of this book.

- Dashan Gao helped with writing Chapters 2 and 3.

- Xueyang Wu helped with writing Chapters 3 and 5.

- Xinle Liang helped with writing Chapters 3 and 9.

- Yunfeng Huang helped with writing Chapters 5 and 8.

- Sheng Wan helped with writing Chapters 6 and 8.

- Xiguang Wei helped with writing Chapter 9.

- Pengwei Xing helped with writing Chapters 8 and 10.

Finally, we thank our family for their understanding and continued support. Without them, the book would not have been possible.

Qiang Yang, Yang Liu, Yong Cheng, Yan Kang, Tianjian Chen, and Han Yu
November 2019, Shenzhen, China

CHAPTER 1

Introduction

1.1 MOTIVATION

We have witnessed the rapid growth of machine learning (ML) technologies in empowering diverse artificial intelligence (AI) applications, such as computer vision, automatic speech recognition, natural language processing, and recommender systems [Pouyanfar et al., 2019, Hatcher and Yu, 2018, Goodfellow et al., 2016]. The success of these ML technologies, in particular deep learning (DL), has been fueled by the availability of vast amounts of data (a.k.a. the big data) [Trask, 2019, Pouyanfar et al., 2019, Hatcher and Yu, 2018]. Using these data, DL systems can perform a variety of tasks that can sometimes exceed human performance; for example, DL empowered face-recognition systems can achieve commercially acceptable levels of performance given millions of training images. These systems typically require a huge amount of data to reach a satisfying level of performance. For example, the object detection system from Facebook has been reported to be trained on 3.5 billion images from Instagram [Hartmann, 2019].

In general, the big data required to empower AI applications is often large in size. However, in many application domains, people have found that big data are hard to come by. What we have most of the time are "small data," where either the data are of small sizes only, or they lack certain important information, such as missing values or missing labels. To provide sufficient labels for data often requires much effort from domain experts. For example, in medical image analysis, doctors are often employed to provide diagnosis based on scan images of patient organs, which is tedious and time consuming. As a result, high-quality and large-volume training data often cannot be obtained. Instead, we face silos of data that cannot be easily bridged.

The modern society is increasingly made aware of issues regarding the data ownership: who has the right to use the data for building AI technologies? In an AI-driven product recommendation service, the service owner claims ownership over the data about the products and purchase transactions, but the ownership over the data about user purchasing behaviors and payment habits is unclear. Since data are generated and owned by different parties and organizations, a traditional and naive approach is to collect and transfer the data to one central location where powerful computers can train and build ML models. Today, this methodology is no longer valid.

While AI is spreading into ever-widening application sectors, concerns regarding user privacy and data confidentiality expand. Users are increasingly concerned that their private information is being used (or even abused) by commercial and political purposes without their permission. Recently, several large Internet corporations have been fined heavily due to their leakage of users' private data to commercial companies. Spammers and under-the-table data exchanges are often punished in court cases.

In the legal front, law makers and regulatory bodies are coming up with new laws ruling how data should be managed and used. One prominent example is the adoption of the General Data Protection Regulation (GDPR) by the European Union (EU) in 2018 [GDPR website, 2018]. In the U.S., the California Consumer Privacy Act (CCPA) will be enacted in 2020 in the state of California [DLA Piper, 2019]. China's Cyber Security Law and the General Provisions of Civil Law, implemented in 2017, also imposed strict controls on data collection and transactions. Appendix A provides more information about these new data protection laws and regulations.

Under this new legislative landscape, collecting and sharing data among different organizations is becoming increasingly difficult, if not outright impossible, as time goes by. In addition, the sensitive nature of certain data (e.g., financial transactions and medical records) prohibits free data circulation and forces the data to exist in isolated data silos maintained by the data owners [Yang et al., 2019]. Due to industry competition, user privacy, data security, and complicated administrative procedures, even data integration between different departments of the *same company faces heavy resistance*. The prohibitively high cost makes it almost impossible to integrate data scattered in different institutions [WeBank AI, 2019]. Now that the old privacy-intrusive way of collecting and sharing data is outlawed, data consolidation involving different data owners is extremely challenging going forward.

How to solve the problem of data fragmentation and isolation while complying with the new stricter privacy-protection laws is a major challenge for AI researchers and practitioners. Failure to adequately address this problem will likely lead to a new AI winter [Yang et al., 2019].

Another reason why the AI industry is facing a data plight is that the benefit of collaborating over the sharing of the big data is not clear. Suppose that two organizations wish to collaborate on medical data in order to train a joint ML model. The traditional method of transferring the data from one organization to another will often mean that the original data owner will lose control over the data that they owned in the first place. The value of the data decreases as soon as the data leaves the door. Furthermore, when the better model as a result of integrating the data sources gained benefit, it is not clear how the benefit is fairly distributed among the participants. This fear of losing control and lack of transparency in determining the distribution of values is causing the so-called data fragmentation to intensify.

With edge computing over the Internet of Things, the big data is often not a single monolithic entity but rather distributed among many parties. For example, satellites taking images of the Earth cannot expect to transmit all data to data centers on the ground, as the amount of transmission required will be too large. Likewise, with autonomous cars, each car must be able to process much information locally with ML models while collaborate globally with other cars and computing centers. How to enable the updating and sharing of models among the multiple sites in a secure and yet efficient way is a new challenge to the current computing methodologies.

1.2 FEDERATED LEARNING AS A SOLUTION

As mentioned previously, multiple reasons make the problem of data silos become impediment to the big data needed to train ML models. It is thus natural to seek solutions to build ML models that do not rely on collecting all data to a centralized storage where model training can happen. An idea is to train a model at each location where a data source resides, and then let the sites communicate their respective models in order to reach a consensus for a global model. In order to ensure user privacy and data confidentiality, the communication process is carefully engineered so that no site can second-guess the private data of any other sites. At the same time, the model is built as if the data sources were combined. This is the idea behind "federated machine learning" or "federated learning" for short.

Federated learning was practiced in an edge-server architecture by McMahan et al. in the context of updating language models on mobile phones [McMahan et al., 2016a,b, Konecný et al., 2016a,b]. There are many mobile edge devices each holding private data. To update the prediction models in the Gboard system, which is the Google's keyboard system for auto-completion of words, researchers at Google developed a federated learning system to update a collective model periodically. Users of the Gboard system gets a suggested query and whether the users clicked the suggested words. The word-prediction model in Gboard keeps improving based on not just a single mobile phone's accumulated data but all phones via a technique known as federated averaging (`FedAvg`). Federated averaging does not require moving data from any edge device to one central location. Instead, with federated learning, the model on each mobile device, which can be a smartphones or a tablet, gets encrypted and shipped to the cloud. All encrypted models are integrated into a global model under encryption, so that the server at the cloud does not know the data on each device [Yang et al., 2019, McMahan et al., 2016a,b, Konecný et al., 2016a,b, Hartmann, 2018, Liu et al., 2019]. The updated model, which is under encryption, is then downloaded to all individual devices on the edge of the cloud system [Konecný et al., 2016b, Hartmann, 2018, Yang et al., 2018, Hard et al., 2018]. In the process, users' individual data on each device is not revealed to others, nor to the servers in the cloud.

Google's federated learning system shows a good example of B2C (business-to-consumer), in designing a secure distributed learning environment for B2C applications. In the B2C setting, federated learning can ensure privacy protection as well as increased performance due to a speedup in transmitting the information between the edge devices and the central server.

Besides the B2C model, federated learning can also support the B2B (business-to-business) model. In federated learning, a fundamental change in algorithmic design methodology is, instead of transferring data from sites to sites, we transfer model parameters in a secure way, so that other parties cannot "second guess" the content of others' data. Below, we give a formal categorization of the federated learning in terms of how the data is distributed among the different parties.

1.2.1   THE DEFINITION OF FEDERATED LEARNING

Federated learning aims to build a joint ML model based on the data located at multiple sites. There are two processes in federated learning: model training and model inference. In the process of model training, information can be exchanged between parties but not the data. The exchange does not reveal any protected private portions of the data at each site. The trained model can reside at one party or shared among multiple parties.

At inference time, the model is applied to a new data instance. For example, in a B2B setting, a federated medical-imaging system may receive a new patient who's diagnosis come from different hospitals. In this case, the parties collaborate in making a prediction. Finally, there should be a fair value-distribution mechanism to share the profit gained by the collaborative model. Mechanism design should done in such a way to make the federation sustainable.

In broad terms, federated learning is an algorithmic framework for building ML models that can be characterized by the following features, where a model is a function mapping a data instance at some party to an outcome.

- There are two or more parties interested in jointly building an ML model. Each party holds some data that it wishes to contribute to training the model.

- In the model-training process, the data held by each party does not leave that party.

- The model can be transferred in part from one party to another under an encryption scheme, such that other parties cannot re-engineer the data at any given party.

- The performance of the resulting model is a good approximation of ideal model built with all data transferred to a single party.

More formally, consider N data owners $\{\mathcal{F}_i\}_{i=1}^{N}$ who wish to train a ML model by using their respective datasets $\{\mathcal{D}_i\}_{i=1}^{N}$. A conventional approach is to collect all data $\{\mathcal{D}_i\}_{i=1}^{N}$ together at one data server and train a ML model $\mathcal{M}_{SUM}$ on the server using the centralized dataset. In the conventional approach, any data owner $\{\mathcal{F}_i$ will expose its data $\{\mathcal{D}_i$ to the server and even other data owners.

Federated learning is a ML process in which the data owners collaboratively train a model $\mathcal{M}_{FED}$ without collecting all data $\{\mathcal{D}_i\}_{i=1}^{N}$. Denote $\mathcal{V}_{SUM}$ and $\mathcal{V}_{FED}$ as the performance measure (e.g., accuracy, recall, and F1-score) of the centralized model $\mathcal{M}_{SUM}$ and the federated model $\mathcal{M}_{FED}$, respectively.

We can capture what we mean by performance guarantee more precisely. Let δ be a non-negative real number. We say that the federated learning model $\mathcal{M}_{FED}$ has δ-performance loss if

$$|\mathcal{V}_{SUM} - \mathcal{V}_{FED}| < \delta. \tag{1.1}$$

The previous equation expresses the following intuition: if we use secure federated learning to build a ML model on distributed data sources, this model's performance on future data is approximately the same as the model that is built on joining all data sources together.

We allow the federated learning system to perform a little less than a joint model because in federated learning data owners do not expose their data to a central server or any other data owners. This additional security and privacy guarantee can be worth a lot more than the loss in accuracy, which is the δ value.

A federated learning system may or may not involve a central coordinating computer depending on the application. An example involving a coordinator in a federated learning architecture is shown in Figure 1.1. In this setting, the coordinator is a central aggregation server (a.k.a. the parameter server), which sends an initial model to the local data owners A–C (a.k.a. clients or participants). The local data owners A–C each train a model using their respective dataset, and send the model weight updates to the aggregation server. The aggregation sever then combines the model updates received from the data owners (e.g., using federated averaging [McMahan et al., 2016a]), and sends the combined model updates back to the local data owners. This procedure is repeated until the model converges or until the maximum number of iterations is reached. Under this architecture, the raw data of the local data owners never leaves the local data owners. This approach not only ensures user privacy and data security, but also saves communication overhead needed to send raw data. The communication between the central aggregation server and the local data owners can be encrypted (e.g., using homomorphic encryption [Yang et al., 2019, Liu et al., 2019]) to guard against information leakage.

The federated learning architecture can also be designed in a peer to peer manner, which does not require a coordinator. This ensures further security guarantee in which the parties communicate directly without the help of a third party, as illustrated in Figure 1.2. The advantage of this architecture is increased security, but a drawback is potentially more computation to encrypt and decrypt messages.

Federated learning brings several benefits. It preserves user privacy and data security by design since no data transfer is required. Federated learning also enables several parties to collaboratively train a ML model so that each of the parties can enjoy a better model than what it can achieve alone. For example, federated learning can be used by private commercial banks to detect multi-party borrowing, which has always been a headache in the banking industry, especially in the Internet finance industry [WeBank AI, 2019]. With federated learning, there is no need to establish a central database, and any financial institution participating in federated learning can initiate new user queries to other agencies within the federation. Other agencies only need to answer questions about local lending without knowing specific information of the user. This not only protects user privacy and data integrity, but also achieves an important business objective of identifying multi-party lending.

While federated learning has great potential, it also faces several challenges. The communication link between the local data owner and the aggregation server may be slow and un-

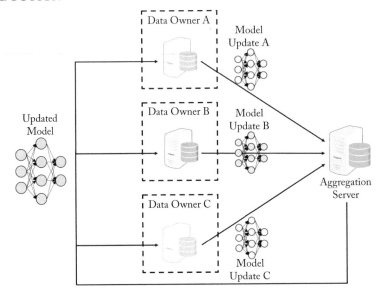

Figure 1.1: An example federated learning architecture: client-server model.

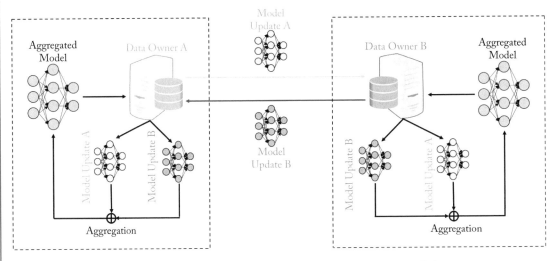

Figure 1.2: An example federated learning architecture: peer-to-peer model.

stable [Hartmann, 2018]. There may be a very large number of local data owners (e.g., mobile users). In theory, every mobile user can participate in federated learning, making the system unstable and unpredictable. Data from different participants in federated learning may follow non-identical distributions [Zhao et al., 2018, Sattler et al., 2019, van Lier, 2018], and different participants may have unbalanced numbers of data samples, which may result in a biased model or even failure of training a model. As the participants are distributed and difficult to authenticate, federated learning model poisoning attacks [Bhagoji et al., 2019, Han, 2019], in which one or more malicious participants send ruinous model updates to make the federated model useless, can take place and confound the whole operation.

1.2.2   CATEGORIES OF FEDERATED LEARNING

Let matrix $\mathcal{D}_i$ denote the data held by the i th data owner. Suppose that each row of the matrix $\mathcal{D}_i$ represents a data sample, and each column represents a specific feature. At the same time, some datasets may also contain label data. We denote the feature space as $\mathcal{X}$, the label space as $\mathcal{Y}$, and we use $\mathcal{I}$ to denote the sample ID space. For example, in the financial field, labels may be users' credit. In the marketing field labels may be the user's purchasing desire. In the education field, $\mathcal{Y}$ may be the students' scores. The feature $\mathcal{X}$, label $\mathcal{Y}$, and sample IDs $\mathcal{I}$ constitute the complete training dataset $(\mathcal{I}, \mathcal{X}, \mathcal{Y})$. The feature and sample spaces of the datasets of the participants may not be identical. We classify federated learning into horizontal federated learning (HFL), vertical federated learning (VFL), and federated transfer learning (FTL), according to how data is partitioned among various parties in the feature and sample spaces. Figures 1.3–1.5 show the three federated learning categories for a two-party scenario [Yang et al., 2019].

HFL refers to the case where the participants in federated learning share overlapping data features, i.e., the data features are aligned across the participants, but they differ in data samples. It resembles the situation that the data is horizontally partitioned inside a tabular view. Hence, we also call HFL as sample-partitioned federated learning, or example-partitioned federated learning [Kairouz et al., 2019]. Different from HFL, VFL applies to the scenario where the participants in federated learning share overlapping data samples, i.e., the data samples are aligned amongst the participants, but they differ in data features. It resembles the situation that data is vertically partitioned inside a tabular view. Thus, we also name VFL as feature-partitioned federated learning. FTL is applicable for the case when there is neither overlapping in data samples nor in features.

For example, when the two parties are two banks that serve two different regional markets, they may share only a handful of users but their data may have very similar feature spaces due to similar business models. That is, with limited overlap in users but large overlap in data features, the two banks can collaborate in building ML models through horizontal federated learning [Yang et al., 2019, Liu et al., 2019].

When two parties providing different services but sharing a large amount of users (e.g., a bank and an e-commerce company), they can collaborate on the different feature spaces that they own, leading to a better ML model for both. That is, with large overlap in users but little overlap in data features, the two companies can collaborate in building ML models through vertical federated learning [Yang et al., 2019, Liu et al., 2019]. Split learning, recently proposed by Gupta and Raskar [2018] and Vepakomma et al. [2019, 2018], is regarded here as a special case of vertical federated learning, which enables vertically federated training of deep neural

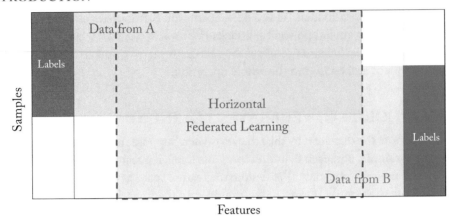

Figure 1.3: Illustration of HFL, a.k.a. sample-partitioned federated learning where the overlapping features from data samples held by different participants are taken to jointly train a model [Yang et al., 2019].

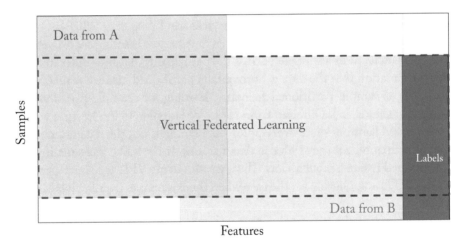

Figure 1.4: Illustration of VFL, a.k.a feature-partitioned federated learning where the overlapping data samples that have non-overlapping or partially overlapping features held by multiple participants are taken to jointly train a model [Yang et al., 2019].

networks (DNNs). That is, split learning facilitates training DNNs in federated learning settings over vertically partitioned data [Vepakomma et al., 2019].

In scenarios where participating parties have highly heterogeneous data (e.g., distribution mismatch, domain shift, limited overlapping samples, and scarce labels), HFL and VFL may not be able to build effective ML models. In those scenarios, we can leverage transfer learning

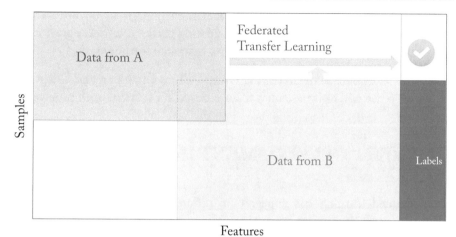

Figure 1.5: Federated transfer learning (FTL) [Yang et al., 2019]. A predictive model learned from feature representations of aligned samples belonging to party A and party B is utilized to predict labels for unlabeled samples of party A.

techniques to bridge the gap between heterogeneous data owned by different parties. We refer to federated learning leveraging transfer learning techniques as FTL.

Transfer learning aims to build effective ML models in a resource-scarce target domain by exploiting or transferring knowledge learned from a resource-rich source domain, which naturally fits the federated learning setting where parties are typically from different domains. Pan and Yang [2010] divides transfer learning into mainly three categories: (i) instance-based transfer, (ii) feature-based transfer, and (iii) model-based transfer. Here, we provide brief descriptions on how these three categories of transfer learning techniques can be applied to federated settings.

- **Instance-based FTL.** Participating parties selectively pick or re-weight their training data samples such that the distance among domain distributions can be minimized, thereby minimizing the objective loss function.

- **Feature-based FTL.** Participating parties collaboratively learn a common feature representation space, in which the distribution and semantic difference among feature representations transformed from raw data can be relieved and such that knowledge can be transferable across different domains to build more robust and accurate shared ML models.

Figure 1.5 illustrates an FTL scenario where a predictive model learned from feature representations of aligned samples belonging to party A and party B is utilized to predict labels for unlabeled samples of party A. We will elaborate on how this FTL is performed in Chapter 6.

- **Model-based FTL.** Participating parties collaboratively learn shared models that can benefit for transfer learning. Alternatively, participating parties can utilize pre-trained models as the whole or part of the initial models for a federated learning task.

We will further explain in detail the HFL and VFL in Chapter 4 and Chapter 5, respectively. In Chapter 6, we will elaborate on a feature-based FTL framework proposed by Liu et al. [2019].

1.3   CURRENT DEVELOPMENT IN FEDERATED LEARNING

The idea of federated learning has appeared in different forms throughout the history of computer science, such as privacy-preserving ML [Fang and Yang, 2008, Mohassel and Zhang, 2017, Vaidya and Clifton, 2004, Xu et al., 2015], privacy-preserving DL [Liu et al., 2016, Phong, 2017, Phong et al., 2018], collaborative ML [Melis et al., 2018], collaborative DL [Zhang et al., 2018, Hitaj et al., 2017], distributed ML [Li et al., 2014, Wang, 2016], distributed DL [Vepakomma et al., 2018, Dean et al., 2012, Ben-Nun and Hoefler, 2018], and federated optimization [Li et al., 2019, Xie et al., 2019], as well as privacy-preserving data analytics [Mangasarian et al., 2008, Mendes and Vilela, 2017, Wild and Mangasarian, 2007, Bogdanov et al., 2014]. Chapters 2 and 3 will present some examples.

1.3.1   RESEARCH ISSUES IN FEDERATED LEARNING

Federated learning was studied by Google in a research paper published in 2016 on arXiv.[1] Since then, it has been an area of active research in the AI community as evidenced by the fast-growing volume of preprints appearing on arXiv. Yang et al. [2019] provide a comprehensive survey of recent advances of federated learning.

Recent research work on federated learning are mainly focused on improving security and statistical challenges [Yang et al., 2019, Mancuso et al., 2019]. Cheng et al. [2019] proposed SecureBoost in the setting of vertical federated learning, which is a novel lossless privacy-preserving tree-boosting system. SecureBoost provides the same level of accuracy as the non-privacy-preserving approach. It is theoretically proven that the SecureBoost framework is as accurate as other non-federated gradient tree-boosting algorithms that rely on centralized datasets [Cheng et al., 2019].

Liu et al. [2019] presents a flexible federated transfer learning framework that can be effectively adapted to various secure multi-party ML tasks. In this framework, the federation allows knowledge to be shared without compromising user privacy, and enables complimentary knowledge to be transferred in the network via transfer learning. As a result, a target-domain party can build more flexible and powerful models by leveraging rich labels from a source-domain party.

---

[1]arXiv is a repository of electronic preprints (e-prints) hosted by Cornell University. For more information, visit arXiv website https://arxiv.org/.

In a federated learning system, we can assume that participating parties are honest, semi-honest, or malicious. When a party is malicious, it is possible for a model to taint its data in training. The possibility of model poisoning attacks on federated learning initiated by a single non-colluding malicious agent is discussed in Bhagoji et al. [2019]. A number of strategies to carry out model poisoning attack were investigated. It was shown that even a highly constrained adversary can carry out model poisoning attacks while simultaneously maintaining stealth. The work of Bhagoji et al. [2019] reveals the vulnerability of the federated learning settings and advocates the need to develop effective defense strategies.

Re-examining the existing ML models under the federated learning settings has become a new research direction. For example, combining federated learning with reinforcement learning has been studied in Zhuo et al. [2019], where Gaussian differentials on the information shared among agents when updating their local models were applied to protect the privacy of data and models. It has been shown that the proposed federated reinforcement learning model performs close to the baselines that directly take all joint information as input [Zhuo et al., 2019].

Another study in Smith et al. [2017] showed that multi-task learning is naturally suited to handle the statistical challenges of federated learning, where separate but related models are learned simultaneously at each node. The practical issues, such as communication cost, stragglers, and fault tolerance in distributed multi-task learning and federated learning, were considered. A novel systems-aware optimization method was put forward, which achieves significant improved efficiency compared to the alternatives.

Federated learning has also been applied in the fields of computer vision (CV), e.g., medical image analysis [Sheller et al., 2018, Liu et al., 2018, Huang and Liu, 2019], natural language processing (NLP) (see, e.g., Chen et al. [2019]), and recommender systems (RS) (see, e.g., Ammad-ud-din et al. [2019]). This will be further reviewed in Chapter 8.

Regarding applications of federated learning, the researchers at Google have applied federated learning in mobile keyboard prediction [Bonawitz and Eichner et al., 2019, Yang et al., 2018, Hard et al., 2018], which has achieved significant improvement in prediction accuracy without exposing mobile user data. Researchers at Firefox have used federated learning for search word prediction [Hartmann, 2018]. There is also new research effort to make federated learning more personalizable [Smith et al., 2017, Chen et al., 2018].

1.3.2   OPEN-SOURCE PROJECTS

Interest in federated learning is not only limited to theoretical work. Research on the development and deployment of federated learning algorithms and systems is also flourishing. There are several fast-growing open-source projects of federated learning.

• **Federated AI Technology Enabler (FATE)** [WeBank FATE, 2019] is an open-source project initiated by the AI department of WeBank[2] to provide a secure computing framework to support the federated AI ecosystem [WeBank FedAI, 2019]. It implements secure computation protocols based on homomorphic encryption (HE) and secure multi-party computation (MPC). It supports a range of federated learning architectures and secure computation algorithms, including logistic regression, tree-based algorithms, DL (artificial neural networks), and transfer learning. For more information on FATE, readers can refer to the GitHub FATE website [WeBank FATE, 2019] and the FedAI website [WeBank FedAI, 2019].

• **TensorFlow[3] Federated** project [Han, 2019, TFF, 2019, Ingerman and Ostrowski, 2019, Tensorflow-federated, 2019] (TFF) is an open-source framework for experimenting with federated ML and other computations on decentralized datasets. TFF enables developers to simulate existing federated learning algorithms on their models and data, as well as to experiment with novel algorithms. The building blocks provided by TFF can also be used to implement non-learning computations, such as aggregated analytics over decentralized data. The interfaces of TFF are organized in two layers: (1) the federated learning (FL) application programming interface (API) and (2) federated Core (FC) API. TFF enables developers to declaratively express federated computations, so that they can be deployed in diverse runtime environments. Included in TFF is a single-machine simulation run-time for experimentation.

• **TensorFlow-Encrypted** [TensorFlow-encrypted, 2019] is a Python library built on top of TensorFlow for researchers and practitioners to experiment with privacy-preserving ML. It provides an interface similar to that of TensorFlow, and aims to make the technology readily available without requiring user to be experts in ML, cryptography, distributed systems, and high-performance computing.

• **coMind** [coMind.org, 2018, coMindOrg, 2019] is an open-source project for training privacy-preserving federated DL models. The key component of coMind is the implementation of the federated averaging algorithm [McMahan et al., 2016a, Yu et al., 2018], which is training ML models in a collaborative way while preserving user privacy and data security. coMind is built on top of TensorFlow and provides high-level APIs for implementing federated learning.

• **Horovod** [Sergeev and Balso, 2018, Horovod, 2019], developed by Uber, is an open-source distributed training framework for DL. It is based on the open message passing interface

---

[2]WeBank, opened in December 2014 upon receiving its banking license in China. It is the first digital-only bank in China. WeBank is devoted to offering individuals and SMEs under-served by the current banking system with a variety of convenient and high-quality financial services. For more information on WeBank, please visit `https://www.webank.com/en/`.

[3]TensorFlow is an open-source DL framework, developed and maintained by Google Inc. TensorFlow is widely used in research and implementation of DL. For more information on TensorFlow, readers can refer to its project website `https://www.tensorflow.org/` and its GitHub website `https://github.com/tensorflow`.

(MPI) and works on top of popular DL frameworks, such as TensorFlow and PyTorch.[4] The goal of Horovod is to make distributed DL fast and easy to use. Horovod supports federated learning via open MPI and currently, encryption is not yet supported.

- **OpenMined/PySyft** [Han, 2019, OpenMined, 2019, Ryffel et al., 2018, PySyft, 2019, Ryffel, 2019] provides two methods for privacy preservation: (1) federated learning and (2) differential privacy. OpenMined further supports two methods of secure computation through multi-party computation and homomorphic encryption. OpenMined has made available the PySyft library [PySyft, 2019], which is an open-source federated learning framework for building secure and scalable ML models [Ryffel, 2019]. PySyft is simply a hooked extension of PyTorch. For users who are familiar with PyTorch, it is very easy to implement federated learning systems with PySyft. Federated learning extension based on the TensorFlow framework is currently being developed within OpenMined.

- **LEAF Beanchmark** [LEAF, 2019, Caldas et al., 2019], maintained by Carnegie Mellon University and Google AI, is a modular benchmarking framework for ML in federated settings, with applications in federated learning, multi-task learning, meta-learning, and on-device learning. LEAF includes a suite of open-source federated datasets (e.g., FEMNIST, Sentiment140, and Shakespeare), a rigorous evaluation framework, and a set of reference implementations, aiming to capture the reality, obstacles, and intricacies of practical federated learning environments. LEAF enables researchers and practitioners in these domains to investigate new proposed solutions under more realistic assumptions and settings. LEAF will include additional tasks and datasets in its future releases.

1.3.3   STANDARDIZATION EFFORTS

As more developments are made in the legal front on the secure and responsible use of users' data, technical standard needs to be developed to ensure that organizations use the same language and follow a standard guideline in developing future federated learning systems. Moreover, there is increasing need for the technical community to communicate with the regulatory and legal communities over the use of the technology. As a result, it is important to develop international standards that can be adopted by multiple disciplines.

For example, companies striving to satisfy the GDPR requirements need to know what technical developments are needed in order to satisfy the legal requirements. Standards can provide a bridge between regulators and technical developers.

One of the early standards is initiated by the AI Department at WeBank with the Institute of Electrical and Electronics Engineers (IEEE) P3652.1 Federated Machine Learning Working Group (known as Federated Machine Learning (C/LT/FML)) was established in December

---

[4]PyTorch is a popular DL framework and is widely used in research and implementation. For more information, visit the official PyTorch website `https://pytorch.org/` and the GitHub PyTorch website `https://github.com/pytorch/pytorch`.

2018 [IEEE P3652.1, 2019]. The objective of this working group is to provide guidelines for building the architectural framework and applications of federated ML. The working group will define the architectural framework and application guidelines for federated ML, including:

1. The description and definition of federated learning;

2. The types of federated learning and the application scenarios to which each type applies;

3. Performance evaluation of federated learning; and

4. The associated regulatory requirements.

The purpose of this standard is to provide a feasible solution for the industrial application of AI without exchanging data directly. This standard is expected to promote and facilitate collaborations in an environment where privacy and data protection issues have become increasingly important. It will promote and enable to the use of distributed data sources for the purpose of developing AI without violating regulations or ethical concerns.

1.3.4   THE FEDERATED AI ECOSYSTEM

The Federated AI (FedAI) ecosystem project was initiated by the AI Department of We-Bank [WeBank FedAI, 2019]. The primary goal of the project is to develop and promote advanced AI technologies that preserve user privacy, data security, and data confidentiality. The federated AI ecosystem features four main themes.

- Open-source technologies: FedAI aims to accelerate open-source development of federated ML and its applications. The FATE project [WeBank FATE, 2019] is a flagship project under FedAI.

- Standards and guidelines: FedAI, together with partners, are drawing up standardization to formulate the architectural framework and application guidelines of federated learning, and facilitate industry collaboration. One representative work is the IEEE P3652.1 federated ML working group [IEEE P3652.1, 2019].

- Multi-party consensus mechanisms: FedAI is studying incentive and reward mechanisms to encourage more institutions to participate in federated learning research and development in a sustainable way. For example, FedAI is undertaking work to establish a multi-party consensus mechanism based on technologies like blockchain.

- Applications in various verticals: To open up the potential of federated learning, FedAI endeavors to showcase more vertical field applications and scenarios, and to build new business models.

1.4   ORGANIZATION OF THIS BOOK

The organization of this book is as follows. Chapter 2 provides background information on privacy-preserving ML, covering well-known techniques for data security. Chapter 3 describes distributed ML, highlighting the difference between federated learning and distributed ML. Horizontal federated learning, vertical federated learning, and federated transfer learning are elaborated in detail in Chapter 4, Chapter 5, and Chapter 6, respectively. Incentive mechanism design for motivating the participation in federated learning is discussed in Chapter 7. Recent work on extending federated learning to the fields of computer vision, natural language processing, and recommender systems are reviewed in Chapter 8. Chapter 9 presents federated reinforcement learning. The prospect of applying federated learning into various industrial sectors is summarized in Chapter 10. Finally, we provide a summary of this book and looking ahead in Chapter 11. Appendix A provides an overview of recent data protection laws and regulations in the European Union, the United States, and China.

CHAPTER 2

Background

In this chapter, we introduce the background knowledge related to federated learning, covering privacy-preserving machine learning techniques and data analytics.

2.1 PRIVACY-PRESERVING MACHINE LEARNING

Data leakage and privacy violation incidents have brought about heightened public awareness of the need for AI systems to be able to preserve user privacy and data confidentiality. Researchers are interested in developing techniques for privacy-preserving properties to be built inside machine learning (ML) systems. The resulting systems are known as privacy-preserving machine learning systems (PPML). In fact, 2018 was considered a breakout year for PPML [Mancuso et al., 2019]. PPML is a broad term that generally refers to ML equipped with defense measures for protecting user privacy and data security. The system security and cryptography community has also proposed various secure frameworks for ML.

In Westin [1968], Westin defined information privacy as follows: "the claim of individuals, groups, or institutions to determine for themselves when, how, and to what extent information about them is communicated to others." This essentially defines the right to control the access and handling of one's information. The main idea of information privacy is to have control over the collection and handling of one's personal data [Mendes and Vilela, 2017].

In this chapter, we will introduce several popular approaches used in PPML including secure multi-party computation (MPC), homomorphic encryption (HE) for privacy-preserving model training and inference, as well as differential privacy (DP) for preventing unwanted data disclosure. Privacy-preserving gradient descent methods will also be discussed.

2.2 PPML AND SECURE ML

Before going into the details of PPML, we first clarify the difference between PPML and secure ML. PPML and secure ML differ mainly in the types of security violations that they are designed to deal with [Barreno et al., 2006]. In secure ML, the adversary (i.e., attacker) is assumed to violate the *integrity and availability* of a data-analytic system, while in PPML, the adversary is assumed to violate the *privacy and confidentiality* of an ML system.

Most of the time, compromise in security is caused by the intentional attack by a third party. We are concerned with three major types of attacks in ML.

- **Integrity attack.** An attack on *integrity* may result in intrusion points being classified as normal (i.e., false negatives) by the ML system.

- **Availability attack.** An attack on *availability* may lead to classification errors (both false negatives and false positives) such that the ML system becomes unusable. This is a broader type of integrity attacks.

- **Confidentiality attack.** An attack on *confidentiality* may result in sensitive information (e.g., training data or model) of an ML system being leaked.

Table 2.1 gives a comparison between PPML and secure ML in terms of security violations, adversary attacks, and defense techniques.

Table 2.1: Comparison between PPML and secure ML

	Security Violations	Adversary Attacks	Defence Techniques
PPML	Privacy Confidentiality	Reconstruction attack Inversion attack Membership-inference attack	Secure multi-party computation Homomorphic encryption Differential privacy
Secure ML	Integrity Availability	Poisoning attack Adversarial attack Oracle attack	Defensive distillation Adversarial training Regularization

In this chapter, we mainly focus on PPML and defense techniques against privacy and confidentiality violations in ML. Interested readers can refer to Barreno et al. [2006] for a more detailed explanation of secure ML.

2.3   THREAT AND SECURITY MODELS

2.3.1   PRIVACY THREAT MODELS

In order to preserve privacy and confidentiality in ML, it is important to understand the possible threat models. In ML tasks, the participants usually take up three different roles: (1) as the input party, e.g., the data owner; (2) as the computation party (e.g., the model builder and inference service provider); and (3) as the result party (e.g., the model querier and user) [Bogdanov et al., 2014].

Attacks on ML may happen in any stage, including data publishing, model training, and model inference. *Attribute-inference attacks* can happen in the data publishing stage, where adversaries may attempt to de-anonymize or target data-record owners for malevolent purposes. The attacks during ML model training are called *reconstruction attacks*, where the computation party aims to reconstruct the raw data of the data providers or to learn more information about the data providers than what the model builders intend to reveal.

For federated learning, reconstruction attacks are the major privacy concerns. In the inference phase of ML models, an adversarial result party may conduct *reconstruction attack*, *model inversion attacks*, or *membership-inference attacks*, using reverse engineering techniques to gain extra information about the model or raw training data.

Reconstruction Attacks. In reconstruction attacks, the adversary's goal is to extract the training data or feature vectors of the training data during ML model training or model inference. In centralized learning, raw data from different data parties are uploaded to the computation party, which makes the data vulnerable to adversaries, such as a malicious computation party. Large companies may collect raw data from users to train an ML model. However, the collected data may be used for other purposes or sent to a third-party without informed consent from the users. In federated learning, each participating party carries out ML model training using their local data. Only the model weight updates or gradient information are shared with other parties. However, the gradient information may also be leveraged to reveal extra information about the training data [Aono et al., 2018]. Plain-text gradient updating may also violate privacy in some application scenarios. To resist reconstruction attacks, ML models that store explicit feature values such as support vector machine (SVM) and k-nearest neighbors (kNN) should be avoided. During model training, secure multi-party computation (MPC) [Yao, 1982] and homomorphic encryption (HE) [Rivest et al., 1978] can be used to defend against such attacks by keeping the intermediate values private. During model inference, the party computing the inference result should only be granted black-box access to the model. MPC and HE can be leveraged to protect the privacy of the user query during model inference. MPC, HE, and their corresponding applications in PPML will be introduced in Sections 2.4.1 and 2.4.2, respectively.

Model Inversion Attacks. In model inversion attacks, the adversary is assumed to have either white-box access or black-box access to the model. For the case of white-box access, the adversary knows the clear-text model without stored feature vectors. For the case of black-box access, the adversary can only query the model with data and collect the responses. The adversary's target is to extract the training data or feature vectors of the training data from the model. The black-box access adversary may also reconstruct the clear-text model from the response by conducting an *equation solving attack*. Theoretically, for an N-dimensional linear model, an adversary can steal it with $N + 1$ queries. Such a problem can be formalized as solving θ from $(x, h_\theta(x))$. The adversary can also learn a similar model using the query-response pairs to simulate the original model. To resist model inversion attacks, less knowledge of the model should be exposed to the adversary. The access to model should be limited to black-box access, and the output should be limited as well. There are several strategies proposed to reduce the success rate of model inversion attack. Fredrikson et al. [2015] choose to report only rounded confidence values. Al-Rubaie and Chang [2016] take the predicted class labels as response, and the aggregated prediction results of multiple testing instances are returned to further enhance model protection. Bayesian neural networks combined with homomorphic encryption have been developed [Xie et al., 2019], to resist such attacks during secure neural network inference.

Membership-Inference Attacks. In membership-inference attacks, the adversary has black-box access to a model, as well as a certain sample, as its knowledge. The adversary's target is to learn if the sample is inside the training set of the model. The adversary infers whether a sample belongs to the training set or not based on the ML model output. The adversary conducts such attacks by finding and leveraging the differences in the model predictions on the samples belonging to the training set vs. other samples. Defense techniques that are proposed to resist model inversion attacks, such as result generalization by reporting rounded prediction results are shown to be effective to thwart such attacks [Shokri et al., 2017]. Differential privacy (DP) [Dwork et al., 2006] is a major approach to resist membership inference attacks, which will be introduced in Section 2.4.3.

Attribute-Inference Attacks. In attribute-inference attacks, the adversary tries to de-anonymize or target record owners for malevolent purpose. Anonymization by removing personally identifiable information (PII) (also known as sensitive features), such as user IDs and names, before data publishing appears to be a natural approach for protecting user privacy. However, it has been shown to be ineffective. For example, Netflix, the world's largest online movie rental service provider, released a movie rating dataset, which contains anonymous movie ratings from 500,000 subscribers. Despite anonymization, Narayanan and Shmatikov [2008] managed to leverage this dataset along with the Internet Movie Database (IMDB) as background knowledge to re-identify the Netflix users in the records, and further managed to deduce the user's apparent political preferences. This incident shows that anonymization fails in the face of strong adversaries with access to alternative background knowledge. To deal with attribute-inference attacks, group anonymization privacy approaches have been proposed in Mendes and Vilela [2017]. Privacy preservation in group anonymization privacy is achieved via generalization and suppression mechanisms.

Model Poisoning Attacks. It has been shown that federated learning may be vulnerable to model poisoning attacks [Bhagoji et al., 2019], also known as backdoor attacks [Bagdasaryan et al., 2019]. For example, a malicious participant in federated learning may inject a hidden backdoor functionality into the trained federated model, e.g., to cause a trained word predictor to complete certain sentences with an attacker-chosen word [Bagdasaryan et al., 2019]. Bhagoji et al. [2019] proposed a number of strategies to carry out model poisoning attacks, such as boosting of the malicious participant's model update, an alternating minimization strategy that alternately optimizes for the legit training loss and the adversarial backdoor objective, and using parameter estimation for the benign updates to improve attack success. Bagdasaryan et al. [2019] developed a new model-poisoning methodology using model replacement, where a constrain-

and-scale technique is proposed to evade anomaly detection-based defenses by incorporating the evasion into the attacker's loss function during model training. Possible solutions against model poisoning attacks include blockchain-based approaches [Preuveneers et al., 2018] and trusted execution environment (TEE) based approaches [Mo and Haddadi, 2019].

2.3.2 ADVERSARY AND SECURITY MODELS

For cryptographic PPML techniques, including MPC and HE, two types of adversaries are concerned in the literature.

- **Semi-honest adversaries.** In the semi-honest (a.k.a honest-but-curious, and passive) adversary model, the adversaries abide by the protocol honestly, but also attempt to learn more information beyond the output from the received information.

- **Malicious adversaries.** In the malicious (a.k.a. active) adversary model, the adversaries deviate from the protocol and can behave arbitrarily.

The semi-honest adversary model is widely considered in most PPML studies. The main reason is that, in federated learning, it is beneficial to each party to honestly follow the ML protocol, since malicious behaviors also break the benefits of the adversaries themselves. The other reason is that, in cryptography, it is a standard method to build a protocol secure against semi-honest adversaries first, then modify it to be secure against malicious adversaries via zero-knowledge proof.

For both security models, the adversaries corrupt a fraction of the parties, and the corrupted parties may collude with each other. The corruption of parties can be static or adaptive. The complexity of an adversary can be either polynomial-time or computational unbounded, corresponding to information-theoretic secure and computational secure, respectively. The security in cryptography is based on the notion of indistinguishability. Interested readers can refer to Lindell [2005] and Lindell and Pinkas [2009] for detailed analysis of adversary and security models.

2.4 PRIVACY PRESERVATION TECHNIQUES

In this section, we discuss privacy preservation techniques. We cover three types of such approaches, namely (1) MPC, (2) HE, and (3) DP.

2.4.1 SECURE MULTI-PARTY COMPUTATION

Secure Multi-Party Computation (MPC), a.k.a. secure function evaluation (SFE), was initially introduced as a secure two-party computation problem (the famous Millionaire's Problem), and generalized in 1986 by Andrew Yao [1986]. In MPC, the objective is to jointly compute a function from the private input by each party, without revealing such inputs to the other parties.

MPC tells us that for any functionality, it is possible to compute it without revealing anything other than the output.

Definition

MPC allows us to compute functions of private input values so that each party learns only the corresponding function output value, but not input values from other parties. For example, given a secret value x that is split into n shares so that a party P_i only knows x_i, all parties can collaboratively compute

$$y_1, \ldots, y_n = f(x_1, \ldots, x_n)$$

so that party P_i learns nothing beyond the output value y_i corresponding to its own input x_i.

The standard approach to prove that an MPC protocol is secure is the *simulation paradigm* [Lindell, 2017]. To prove an MPC protocol is secure against adversaries that corrupt t parties under the simulation paradigm, we build a simulator that, when given inputs and outputs of t colluding parties, generates t transcripts, so that the generated transcripts are *indistinguishable* to that generated in the actual protocol.

In general, MPC can be implemented through three different frameworks, namely: (1) Oblivious Transfer (OT) [Keller et al., 2016, Goldreich et al., 1987]; (2) Secret Sharing (SS) [Shamir, 1979, Rabin and Ben-Or, 1989]; and (3) Threshold Homomorphic Encryption (THE) [Cramer et al., 2001, Damgård and Nielsen, 2003]. From a certain point of view, both oblivious transfer protocols and threshold homomorphic encryption schemes use the idea of secret sharing. This might be the reason why secret sharing is widely regarded as the core of MPC. In the rest of this section, we will introduce oblivious transfer and secret sharing.

Oblivious Transfer

OT is a two-party computation protocol proposed by Rabin in 1981 [Rabin, 2005]. In OT, the sender owns a database of message-index pairs $(M_1, 1), \ldots, (M_N, N)$. At each transfer, the receiver chooses an index i for some $1 \leq i \leq N$, and receives M_i. The receiver does not learn any other information about the database, and the sender does not learn anything about the receiver's selection i. Here, we give the definition of 1-out-of-n OT.

Definition 2.1   1-out-of-n OT: Suppose Party A has a list $(x_1, \ldots, x_n)$ as the input, Party B has $i \in 1, \ldots, n$ as the input. 1-out-of-n OT is an MPC protocol where A learns nothing about i and B learns nothing else but x_i.

When $n = 2$, we get 1-out-of-2 OT which has the following property: 1-out-of-2 OT is universal for two-party MPC [Ishai et al., 2008]. That is, given a 1-out-of-2 OT protocol, one can conduct any secure two-party computation.

Many Constructions of OT has been proposed such as Bellare–Micali's [Bellare and Micali, 1990], Naor–Pinka's [Naor and Pinkas, 2001], and Hazay–Lindell's [Hazay and Lindell,

2010] approaches. Here, we demonstrate the Bellare-Micali's construction of OT, which utilizes Diffie–Hellman key exchange and is based on the computational Diffie–Hellman (CDH) assumption [Diffie and Hellman, 1976]. The Bellare–Micali's construction works as follows: the receiver sends two public keys to the sender. The receiver only holds one private key corresponding to one of the two public keys, and the sender does not know which public key it is. Then, the sender encrypts the two massages with their corresponding public keys, and sends the ciphertexts to the receiver. Finally, the receiver decrypts the target ciphertext with the private key.

Bellare–Micali Construction. In a discrete logarithm setting $(\mathbb{G}, g, p)$, where $\mathbb{G}$ is a group of prime order p, $g \in \mathbb{G}$ is a generator, and $H : G \rightarrow \{0, 1\}^n$ is a hash function. Suppose the sender A has $x_0, x_1 \in \{0, 1\}^n$, and the receiver B has $b \in \{0, 1\}$.

1. A chooses a random element $c \leftarrow G$ and sends it to B.

2. B chooses $k \leftarrow \mathbb{Z}_p$ and sets $PK_b = g^k$, $PK_{1-b} = c/PK_b$, then sends PK_0 to A. A sets $PK_1 = c/PK_0$.

3. A encrypts x_0 with ElGamal scheme using PK_0 (i.e., setting $C_0 = [g^{r_0}, HASH(PK_0^{r_0}) * x_0]$ and encrypting x_1 using PK_1). Then, A sends (C_0, C_1) to B.

4. B decrypts C_b using private key k to obtain $x_b = PK_b^{r_b} * x_b/g^{r_b k}$.

Yao's Garbled Circuit (GC). [Yao, 1986] is a well-known OT-based secure two-party computation protocol that can evaluate any function. The key idea of Yao's GC is to decompose the computational circuits into generation and evaluation stages. The circuits consisting of gates like AND, OR, and NOT can be used to compute any arithmetic operation. Each party is in charge of one stage and the circuit is garbled in each stage, so that any of them cannot get information from the other one, but they can still achieve the result according to the circuit. GC consists of an OT protocol and a block cipher. The complexity of the circuit grows at least linearly with the input size. Soon after GC was proposed, GMW [Goldreich et al., 1987] extended GC to the multi-party setting against malicious adversaries. For more detailed survey of GC, readers can refer to Yakoubov [2017].

OT Extension. Impagliazzo and Rudich [1989] showed that OT provably requires "public-key" type of assumptions (such as factoring, discrete log, etc.). However, Beaver [1996] pointed out that OT can be "extended" in the sense that it is enough to generate a few "seed" OTs based on public-key cryptography, which can then be extended to any number of OTs using symmetric-key cryptosystems only. OT extension is now widely applied in MPC protocols [Keller et al., 2016, Mohassel and Zhang, 2017, Demmler et al., 2015] to improve efficiency.

Secret Sharing

Secret sharing is a concept of hiding a secret value by splitting it into random parts and distributing these parts (a.k.a. shares) to different parties, so that each party has only one share and thus only one piece of the secret [Shamir, 1979, Beimel, 2011]. Depending on the specific secret sharing schemes used, all or a known threshold of shares are needed to reconstruct the original secret value [Shamir, 1979, Tutdere and Uzunko, 2015]. For example, Shamir's Secret Sharing is constructed based on polynomial equations and provides information-theoretic security, and it is also efficient using matrix calculation speedup [Shamir, 1979]. There are several types of secret sharing, mainly including arithmetic secret sharing [Damård et al., 2011], Shamir's secret sharing [Shamir, 1979], and binary secret sharing [Wang et al., 2007]. As arithmetic secret sharing is mostly adopted by existing SMPC-based PPML approaches and binary secret sharing are closely related to OT which is discussed in Section 2.4.1, here we focus on arithmetic secret sharing.

Consider that a party P_i wants to share a secret S among n parties $\{P_i\}_{i=1}^n$ in a finite field F_q. To share S, the party P_i randomly samples $n-1$ values $\{s_i\}_{i=1}^{n-1}$ from $\mathbb{Z}_q$ and set $s_n = S - \sum_{i=1}^{n-1} s_i \bmod q$. Then, P_i distributes s_k to party P_k, for $k \neq i$. We denote the shared S as $\langle S \rangle = \{s_i\}_{i=1}^n$.

The arithmetic addition operation is carried out locally at each party. The secure multiplication is performed by using Beaver triples [Beaver, 1991]. The Beaver triples can be generated in an offline phase. The offline phase (i.e., preprocessing) serves as a *trusted dealer* who generates Beaver triples $\{(\langle a \rangle, \langle b \rangle, \langle c \rangle)|ab = c\}$ and distributes the shares among the n parties.

To compute $\langle z \rangle = \langle x \rangle \cdot \langle y \rangle = \langle x * y \rangle$, $P_{i\ i=1}^n$ first computes $\langle e \rangle = \langle x \rangle - \langle a \rangle$, $\langle f \rangle = \langle y \rangle - \langle b \rangle$. Then, e and f are reconstructed. Finally, P_i computes $\langle z \rangle = \langle c \rangle + e\langle x \rangle + f\langle y \rangle$ locally, and a random party P_j adds its share into ef. We denote element-wise multiplication of vectors as $\langle \cdot \rangle \odot \langle \cdot \rangle$.

Secure multiplication can also be performed by leveraging the Gilboa's protocol [Gilboa, 1999], in which n-bit arithmetic multiplication can be conducted via n 1-out-of-2 OTs. Suppose that Party A holds x and Party B holds y. Now we show Gilboa's protocol, which results in A holding $\langle z \rangle_A$ and B holding $\langle z \rangle_B$ such that $z = x \cdot y$. Let l be the maximum length of the binary representation of the numbers involved in our protocol. Denote the $m \times$ 1-out-of-2 OT for l-bit strings as OT_l^m. Denote the ith bit of x as $x[i]$. The secure 2-party multiplication via OT can be conducted as follows.

1. A represents x in binary format.

2. B builds OT_l^l. For the ith OT, randomly pick $a_{i,0}$ and compute $a_{i,1} = 2^i y - a_{i,0}$. Use $(-a_{i,0}, a_{i,1})$ as the input for the ith OT.

3. A inputs X[i] as the choice bit in the ith OT and obtains $x[i] \times 2^i y - a_{i,0}$.

4. A computes $\langle z \rangle_A = \sum_{i=1}^l (x[i] \times 2^i y - a_{i,0})$
   B computes $\langle z \rangle_B = \sum_{i=1}^l a_{i,0}$.

The offline phase can be carried out efficiently with the help of a *semi-honest dealer* who generates Beaver triples and distributes them among all the parties. To perform such a preprocessing step without a *semi-honest dealer*, there are several protocols available, such as SPDZ [Damård et al., 2011], SPDZ-2 [Damård et al., 2012], MASCOT [Keller et al., 2016], and HighGear [Keller et al., 2018].

- SPDZ is an offline protocol in the preprocessing model based on somewhat homomorphic encryption (SHE) in the form of BGV, first described in Damård et al. [2011].

- SPDZ-2 [Damård et al., 2012] is a protocol based on threshold SHE cryptography (with a shared decryption key).

- MASCOT is an oblivious-transfer-based protocol, proposed in Keller et al. [2016]. It is far more computationally efficient than SPDZ and SPDZ-2.

- In 2018, Keller et al. [2018] developed a BGV-based SHE protocol, called the HighGear protocol, which achieves better performance than the MASCOT protocol.

Application in PPML

Various MPC-based approaches have been designed and implemented for PPML in the past. Most MPC-based PPML approaches leverage a two-phase architecture, comprising of an offline phase and an online phase. The majority of cryptographic operations are conducted in the offline phase, where multiplication triples are generated. The ML model is then trained in the online phase using the multiplication triples generated in the offline phase. The DeepSecure [Rouhani et al., 2017] is a GC-based framework for secure neural network inference, where the inference function has to be represented as a Boolean circuit. The computation and communication cost in GC only depend on the total number of AND gates in the circuit.

SecureML [Mohassel and Zhang, 2017] is another two-party framework for PPML employing two-phase architecture. Parties in federated learning distributes arithmetic shared of their data among two non-colluding servers, who run secure two-party model training protocols. Both Linearly HE (LHE)-based and OT-based protocols are proposed for multiplication triples generation in offline phase. The online phase is based on arithmetic secret sharing and division GC. Therefore, only linear operations are allowed in model training, and various approximations are done to nonlinear functions.

The Chameleon framework is another hybrid MPC framework based on ABY for neural network model inference [Demmler et al., 2015]. Arithmetic secret sharing is used to conduct linear operations, and GC as well as GMW [Goldreich et al., 1987] are used for nonlinear operations. Conversion protocols are also implemented to convert data representations among different protocols.

Privacy-preserving ID3 learning based on OT is addressed in Lindell and Pinkas [2002]. Shamir's threshold secret sharing is used for secure model aggregation for PPML with security against both honest-but-curious and malicious adversaries [Bonawitz et al., 2017], where a group of clients do MPC to evaluate the average of their private input models, and disclose the average to the parameter server for model update. Recently, MPC-based approaches pursuing security against malicious corrupted majority has been studied. For example, linear regression and logistic regression training and evaluation with SPDZ is studied in Chen et al. [2019]. The authors in Damgård et al. [2019] embraces $SPDZ_{2^k}$ [Cramer et al., 2018] for actively secure private ML against a dishonest majority. It implements decision tree and SVM evaluation algorithms.

2.4.2 HOMOMORPHIC ENCRYPTION

HE is generally considered as an alternative approach to MPC in PPML. HE can also be used to achieve MPC as discussed in Section 2.4.1. The concept of HE was proposed in 1978 by Rivest et al. [1978] as a solution to perform computation over ciphertext without decrypting the ciphertext. Since then, numerous attempts have been made by researchers all over the world to design such homomorphic schemes.

The encryption system proposed by Goldwasser and Micali [1982] was a provably secure encryption scheme that reached a remarkable level of safety. It allows an additive operation over ciphertext, but is able to encrypt only a single bit. Paillier [1999] proposed a provable security encryption system that also allows an additive operation over ciphertext in 1999. It has been widely used in various applications. A few years later, in 2005, Boneh et al. [2005] invented a system of provable security encryption, which allows unlimited number of additive operations and one multiplicative operation. Gentry made a breakthrough in 2009 and proposed the first HE scheme that supports both additive and multiplicative operations for unlimited number of times [Gentry, 2009].

Definition

An HE scheme $\mathcal{H}$ is an encryption scheme that allows certain algebraic operations to be carried out on the encrypted content, by applying an efficient operation to the corresponding ciphertext (without knowing the decryption key). An HE scheme $\mathcal{H}$ consists of a set of four functions:

$$\mathcal{H} = \{KeyGen, Enc, Dec, Eval\}, \tag{2.1}$$

where

- *KeyGen*: Key generation. A cryptographic generator g is taken as the input. For asymmetric HE, a pair of keys $\{pk, sk\} = KeyGen(g)$ are generated, where pk is the public key for encryption of the plaintext and sk is the secret (private) key for decryption of the ciphertext. For symmetric HE, only a secret key $sk = KeyGen(g)$ is generated.

- *Enc*: Encryption. For asymmetric HE, an encryption scheme takes the public key pk and the plaintext m as the input, and generates the ciphertext $c = Enc_{pk}(m)$ as the output. For symmetric HE, an HE scheme takes the secret key sk and the plaintext m, and generates ciphertext $c = Enc_{sk}(m)$.

- *Dec*: Decryption. For both symmetric and asymmetric HE, the secret key *sk* and the ciphertext *c* are taken as the input to produce the corresponding plaintext $m = Dec_{sk}(c)$.

- *Eval*: Evaluation. The function *Eval* takes the ciphertext *c* and the public key *pk* (for asymmetric HE) as the input, and outputs a ciphertext corresponding to a functioned plaintext.

Let $Enc_{enk}(\cdot)$ denote the encryption function with *enk* as the encryption key. Let $\mathcal{M}$ denote the plaintext space and $\mathcal{C}$ denote the ciphertext space. A secure cryptosystem is called *homomorphic* if it satisfies the following condition:

$$\forall m_1, m_2 \in \mathcal{M}, \quad Enc_{enk}(m_1 \odot_{\mathcal{M}} m_2) \leftarrow Enc_{enk}(m_1) \odot_{\mathcal{C}} Enc_{enk}(m_2) \qquad (2.2)$$

for some operators $\odot_{\mathcal{M}}$ in $\mathcal{M}$ and $\odot_{\mathcal{C}}$ in $\mathcal{C}$, where $\leftarrow$ indicates the left-hand side term is equal to or can be directly computed from the right-hand side term without any intermediate decryption. In this book we denote homomorphic encryption operator as $[[\cdot]]$, and we overload the addition and multiplication operators over ciphertexts as follows.

- **Addition:** $Dec_{sk}([[u]] \odot_{\mathcal{C}} [[v]]) = Dec_{sk}([[u+v]])$, where "$\odot_{\mathcal{C}}$" may represent multiplication of the ciphertexts (see, e.g., Paillier [1999]).

- **Scalar multiplication:** $Dec_{sk}([[u]] \odot_{\mathcal{C}} n) = Dec_{sk}([[u \cdot n]])$, where "$\odot_{\mathcal{C}}$" may represent taking the power of *n* of the ciphertext (see, e.g., Paillier [1999]).

Categorization of HE Schemes

HE schemes can be categorized into three classes: Partially Homomorphic Encryption (PHE), Somewhat Homomorphic Encryption (SHE), and Fully Homomorphic Encryption (FHE). In general, for HE schemes, the computational complexity increases as the functionality grows. Here, we provide a brief introduction to different types of HE schemes. Interested readers can refer to Armknecht et al. [2015] and Acar et al. [2018] for more details regarding different classes of HE schemes.

Partially Homomorphic Encryption (PHE). For PHE, both $(\mathcal{M}, \odot_{\mathcal{M}})$ and $(\mathcal{C}, \odot_{\mathcal{C}})$ are groups. The operator $\odot_{\mathcal{C}}$ can be applied on ciphertexts for a unlimited number of times. PHE is a *group homomorphism* technique. Specifically, if $\odot_{\mathcal{M}}$ is addition operator, the scheme is *additively homomorphic*, and if $\odot_{\mathcal{M}}$ is a multiplication operator, we say that the scheme is *multiplicative homomorphic*. The references Rivest et al. [1978] and ElGamal [1985] represent two typical multiplicative HE schemes. Examples of additive HE schemes can be found in Goldwasser and Micali [1982] and Paillier [1999].

Somewhat Homomorphic Encryption (SHE). An HE scheme is called SHE if some operations (e.g., addition and multiplication) can be applied for only a limited number of times. Some literature also refer to the schemes supporting arbitrary operations while only some limited circuits (e.g., the branching programs [Ishai and Paskin, 2007], garbled circuit [Yao, 1982]) as SHE. Examples are BV [Brakerski and Vaikuntanathan, 2011], BGN [Boneh et al., 2005], and IP [Ishai and Paskin, 2007]. SHE schemes introduce *noise* for security. Each operation on the ciphertext increases the noise of the output ciphertext, and multiplication is the main technique for increasing noise. When the noise exceeds an upper bound, decryption cannot be conducted correctly. This is the reason why most SHE schemes require a limited number of times of applying the operations.

Fully Homomorphic Encryption (FHE). FHE schemes allow both additive and multiplicative operations with unlimited number of times over ciphertexts. It is worth noticing that *additive* and *multiplicative* operations are the only two operations needed to compute arbitrary functions. Consider $A, B \in \mathbb{F}_2$. The *NAND* gate can be constructed by $1 + A * B$. Thanks to its functional completeness, the NAND gate can be used to construct any gate. Therefore, any functionality can be evaluated by FHE. There are four main families of FHE [Acar et al., 2018]: (1) Ideal Lattice-based FHE (see, e.g., Gentry [2009]); (2) Approximate-GCD based FHE (see, e.g., Dijk et al. [2010]); (3) (R)LWE-based FHE (e.g., Lyubashevsky et al. [2010] and Brakerski et al. [2011]); and (4) NTRU-like FHE (see, e.g., López-Alt et al. [2012]).

The existing FHE schemes are built on SHE, by assuming circular security and implementing an expensive *bootstrap* operation. The bootstrap operation re-encrypts the ciphertexts, by evaluating the decryption and encryption functions over the ciphertexts and the encrypted secret key, in order to reduce the noise of ciphertext for further computation. As a result of the costly bootstrap operation, FHE schemes are very slow and not competitive against general MPC approaches in practice. Researchers are now focusing on finding more efficient SHE schemes that satisfy certain requirements, instead of trying to develop an FHE scheme. In addition, FHE schemes assume circular security (a.k.a. key dependent message (KDM) security), which keeps the secret key secure by encrypting it with the public key. However, no FHE is proven to be semantically secure with respect to any function and is IND-CCA1 secure [Acar et al., 2018].

Application in PPML

Many research efforts based on HE have been devoted to PPML in the past. For example, Hardy et al. [2017] proposed algorithms for privacy-preserving two-party logistic regression for vertically partitioned data. Paillier's scheme is leveraged in secure gradient descent to train the logistic regression model, where constant-multiplication and addition operations are conducted via a mask encrypted by Paillier's scheme and the intermediate data computed by each party. The encrypted masked intermediate results are exchanged between the two parties in the secure gradient descent algorithm. Finally, the encrypted gradient is sent to a coordinator for decryption and model update.

CryptoNets [Gilad-Bachrach et al., 2016] is an HE-based methodology announced by Microsoft Research that allows secure evaluation (inference) of encrypted queries over already trained neural networks on cloud servers: queries from the clients can be classified securely by the trained neural network model on a cloud server without inferring any information about the query or the result. The CryptoDL [Hesamifard et al., 2017] framework is a leveled HE-based approach for secure neural network inference. In CryptoDL, several activation functions are approximated using low-degree polynomials and mean-pooling is used as a replacement for max-pooling. The GAZELLE [Juvekar et al., 2018] framework is proposed as a scalable and low-latency system for secure neural network inference. In GAZELLE, to conduct secure nonlinear function evaluation in neural network inference, HE and traditional secure two-party computation techniques (such as GC) are combined in an intricate way. The packed additive homomorphic encryption (PAHE) embraced in GAZELLE allows single instruction multiple data (SIMD) arithmetic homomorphic operations over encrypted data.

FedMF [Chai et al., 2019] uses Paillier's HE for secure federated matrix factorization assuming honest-but-curious server and honest clients. Secure federated transfer learning is studied via Paillier's HE scheme in Liu et al. [2019], where the semi-honest third party is into the discard by mixing HE with additive secret sharing in decryption process.

2.4.3   DIFFERENTIAL PRIVACY

DP was originally developed to facilitate secure analysis over sensitive data. With the rise of ML, DP has become an active research field again in the ML community. This is motivated by the fact that many exciting results from DP can be applied to PPML [Dwork et al., 2016, 2006]. The key idea of DP is to confuse the adversaries when they are trying to query individual information from the database so that adversaries cannot distinguish individual-level sensitivity from the query result.

Definition

DP is a privacy definition initially proposed by Dwork et al. [2006], developed in the context of statistical disclosure control. It provides an information-theoretic security guarantee that the outcome of a function to be insensitive to any particular record in the dataset. Therefore, DP can be used to resist the membership inference attack. The (ϵ, δ)-differential privacy is defined as follows.

Definition 2.2 (ϵ, δ)-differential privacy. A randomized mechanism $\mathcal{M}$ preserves (ϵ, δ)-differential privacy if given any two datasets D and D' differing by only one record, and for all $S \subset Range(\mathcal{M})$,

$$\Pr[\mathcal{M}(d) \in S] \leq \Pr[\mathcal{M}(D') \in S] \times e^{\epsilon} + \delta, \tag{2.3}$$

where ϵ is the privacy budget and δ is the failure probability.

The quantity $\ln \frac{\Pr[\mathcal{M}(D) \in S]}{\Pr[\mathcal{M}(D') \in S]}$ is called the *privacy loss*, with ln denoting natural logarithm operation. When $\delta = 0$, the stronger notion of ϵ-differential privacy is achieved.

DP has utility-privacy trade-offs as it introduces noise to data. Jayaraman and Evans [2019] found out that current mechanisms for differential privacy for ML rarely offer acceptable utility-privacy trade-offs: settings that provide limited accuracy loss provide little effective privacy, and settings that provide strong privacy result in large accuracy loss.

Categorization of DP Schemes

Typically, there are mainly two ways to achieve DP by adding noise to the data. One is the addition of noise according to the sensitivity of a function [Dwork et al., 2006]. The other is choosing noise according to an exponential distribution among discrete values [McSherry and Talwar, 2007].

The sensitivity of a real-valued function expresses the maximum possible change in its value due to the addition or removal of a single sample.

Definition 2.3   Sensitivity. For two datasets D and D' differing by only one record, and a function $\mathcal{M} : \mathcal{D} \rightarrow \mathcal{R}^d$ over an arbitrary domain, the sensitivity of $\mathcal{M}$ is the maximum change in the output of $\mathcal{M}$ over all possible inputs:

$$\Delta \mathcal{M} = \max_{D, D'} \| \mathcal{M}(D) - \mathcal{M}(D') \|, \tag{2.4}$$

where $\|\cdot\|$ is a norm of the vector. The l_1-sensitivity or the l_2-sensitivity is defined when the l_1-norm or l_2-norm is applied, respectively.

We denote the Laplace distribution with parameter b as $Lap(b)$. $Lap(b)$ has a probability density function $P(z|b) = \frac{1}{2b} \exp(-|z|/b)$. Given a function $\mathcal{M}$ with sensitivity $\Delta \mathcal{M}$, the addition of noise drawn from a calibrated Laplace distribution $Lap(\Delta \mathcal{M}/\epsilon)$ maintains ϵ-differential privacy [Dwork et al., 2006].

Theorem 2.4   *Given a function $\mathcal{M} : \mathcal{D} \rightarrow \mathcal{R}^d$ over an arbitrary domain D, for any input X, the function:*

$$\mathcal{M}(X) + Lap\left(\frac{\Delta \mathcal{M}}{\epsilon}\right)^d \tag{2.5}$$

provides ϵ-differential privacy. The ϵ-differential privacy can also be achieved by adding independently generated Laplace noise from distribution $Lap(\Delta \mathcal{M}/\epsilon)$ to each of the d output terms.

The addition of Gaussian or binomial noise, scaled to the l_2-sensitivity of the function, sometimes yields better accuracy, while only ensuring the weaker (ϵ, δ)-differential privacy [Dwork et al., 2006, Dwork and Nissim, 2004].

The *exponential mechanism* [McSherry and Talwar, 2007] is another way to obtain DP. The exponential mechanism is given a quality function q that scores outcomes of a calculation, where higher scores are better. For a given database and ϵ parameter, the quality function induces a probability distribution over the output domain, from which the exponential mechanism samples the outcome. This probability distribution favors high-scoring outcomes, while ensuring ϵ-differential privacy.

Definition 2.5   Let $q : (\mathcal{D}^n \times \mathcal{R}) \to \mathbb{R}$ be a quality function, which given a dataset $d \in \mathcal{D}^n$, assigns a score to each outcome $r \in \mathcal{R}$. For any two datasets D and D' differing by only one record, let $S(q) = \max_{r,D,D'} \|q(D,r) - q(D',r)\|_1$. Let $\mathcal{M}$ be a mechanism for choosing an outcome $r \in \mathcal{R}$ given a dataset instance $d \in D^n$. Then, the mechanism $\mathcal{M}$, defined as

$$\mathcal{M}(d,q) = \left\{ \text{return } r \text{ with probability } \propto \exp\left(\frac{\epsilon q(d,r)}{2S(q)}\right) \right\} \qquad (2.6)$$

provides ϵ-differential privacy.

The DP algorithms can be categorized according to how and where the perturbation is applied.

1. **Input perturbation**: The noise is added to the training data.

2. **Objective perturbation**: The noise is added to the objective function of the learning algorithms.

3. **Algorithm perturbation**: The noise is added to the intermediate values such as gradients in iterative algorithms.

4. **Output perturbation**: The noise is added to the output parameters after training.

DP still exposes the statistics of a party, which are sensitive in some cases, such as financial data, medical data and other commercial and health applications. Readers who are interested in DP and willing to learn more about it can refer to the tutorial given by Dwork and Roth [2014].

Application in PPML
In federated learning, to enable model training on distributed datasets held by multiple parties, *local differential privacy* (LDP) can be used. With local differential privacy, each input party would perturb their data, then release the obfuscated data to the un-trusted server. The main idea behind local differential privacy is *randomized response* (RR).

Papernot et al. [2016] utilized the teacher ensemble framework to first learn a teacher model ensemble from the distributed datasets among all the parties. Then, the teacher model ensemble is used to make noisy predictions on a public dataset. Finally, the labeled public dataset is used to train a student model. The privacy loss is precisely controlled by the number of public data samples inferred by the teacher ensemble. Generative adversarial network (GAN) is further applied in Papernot et al. [2018] to generate synthetic training data for the training of the student model. Although this approach is not limited to a single ML algorithm, it requires adequate data quantity at each location.

Moments accountant is proposed for differentially private stochastic gradient descent (SGD), which computes the overall privacy cost in neural networks model training by taking into account the particular noise distribution under consideration [Abadi et al., 2016]. It proves less privacy loss for appropriately chosen settings of the noise scale and the clipping threshold.

The differentially private Long Short Term Memory (LSTM) language model is built with user-level differential privacy guarantees with only a negligible cost in predictive accuracy [McMahan et al., 2017]. Phan et al. [2017] proposed a private convolutional deep belief network (pCDBN) by leveraging the functional mechanism to perturb the energy-based objective functions of traditional convolutional deep belief networks. Generating differentially private datasets using GANs is explored in Triastcyn and Faltings [2018], where a Gaussian noise layer is added to the discriminator of a GAN to make the output and the gradients differentially private with respect to the training data. Finally, the privacy-preserving artificial dataset is synthesized by the generator. In addition to the DP dataset publishing, differentially private model publishing for deep learning is also addressed in Yu et al. [2019], where concentrated DP and a dynamic privacy budget allocator are embraced to improve the model accuracy.

Geyer et al. [2018] studied differentially private federated learning and proposed an algorithm for client-level DP preserving federated optimization. It was shown that DP on a client level is feasible and high model accuracy can be reached when sufficiently many participants are involved in federated learning.

CHAPTER 3

Distributed Machine Learning

As we know from Chapter 1, federated learning and distributed machine learning (DML) share several common features, e.g., both employing decentralized datasets and distributed training. Federated learning is even regarded as a special type of DML by some researchers, see, e.g., Phong and Phuong [2019], Yu et al. [2018], Konecný et al. [2016b] and Li et al. [2019], or seen as the future and the next step of DML. In order to gain deeper insights into federated learning, in this chapter, we provide an overview of DML, covering both the scalability-motivated and the privacy-motivated paradigms.

DML covers many aspects, including distributed storage of training data, distributed operation of computing tasks, and distributed serving of model results, etc. There exist a large volume of survey papers, books, and book chapters on DML, such as Feunteun [2019], Ben-Nun and Hoefler [2018], Galakatos et al. [2018], Bekkerman et al. [2012], Liu et al. [2018], and Chen et al. [2017]. Hence, we do not intend to provide another comprehensive survey on this topic. We focus here on the aspects of DML that are most relevant to federated learning, and refer the readers to the references for more details.

3.1 INTRODUCTION TO DML

3.1.1 THE DEFINITION OF DML

DML, also known as distributed learning, refers to multi-node machine learning (ML) or deep learning (DL) algorithms and systems that are designed to improve performance, preserve privacy, and scale to more training data and bigger models [Trask, 2019, Liu et al., 2017, Galakatos et al., 2018]. For example, as illustrated in Figure 3.1, a DML system with three workers (a.k.a. computing nodes) and one parameter server [Li et al., 2014], the training data are split into disjoint data shards and sent to the workers, and the workers carry out stochastic gradient descent (SGD) at their locality. The workers send gradients $\triangle \mathbf{w}^i$ or model weights $\mathbf{w}^i$ to the parameter server, where the gradients or model weights are aggregated (e.g., via taking weighted average) to obtain the global gradients $\triangle \mathbf{w}$ or model weights $\mathbf{w}$. Both synchronous and asynchronous SGD algorithms can be applied in DML [Ben-Nun and Hoefler, 2018, Chen et al., 2017].

DML can generally be categorized into two classes, namely the scalability-motivated DML and the privacy-motivated DML. The scalability-motivated DML refers to the DML paradigm that is designed to address the ever-increasing scalability and computation requirements of large-scale ML systems. For example, in the past decades, the scales of the problems that ML and DL methods have to deal with have increased exponentially. Training a sophisti-

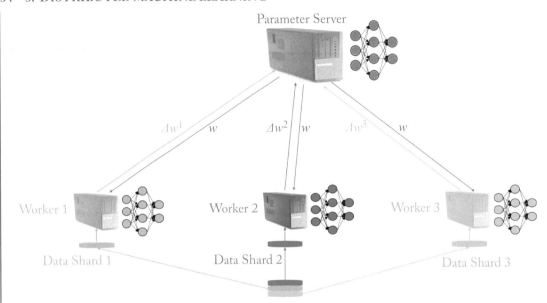

Figure 3.1: Illustration of a distributed machine learning (DML) system.

cated DL model with a huge amount of data can easily exceed the capability of the traditional ML paradigm that relies on a single computing entity. One outstanding example is the famous BERT model [Devlin et al., 2019], which requires multiple tensor processing units (TPUs) for pre-training and it may take several days even with a fleet of TPUs. To cope with such scenarios, the fast-developing DML methods are considered as the answer to the ever-increasing size and complexity of ML models.

Scalability-motivated DML approaches provide feasible solutions to large-scale ML systems when memory limitation and algorithm complexity are the main obstacles. Besides overcoming the problem of requiring centralized storage of training data, DML system can have more elastic and scalable computing resources, such as adding more computing entities on-demand. This is particularly helpful in the age of cloud computing, where we can ask for more processors (such as CPUs, GPUs, or even TPUs) and memory in an on-demand manner. In light of this feature, the scalability-motivated DML is widely applied in the scenarios with horizontally partitioned datasets, where disjoint subsets of training data are stored at different computing entities.

Different from the scalability-motivated DML, the primary goal of privacy-motivated DML paradigm is to preserve user privacy. As user privacy and data security become a global concern (see also Chapter 1 and Appendix A) [Mancuso et al., 2019], privacy-preserving ML is becoming a new trend in the ML community (see also Chapter 2) [Yang et al., 2019]. In a privacy-motivated DML system, there are multiple parties and each of them holds a subset of the training data. Due to privacy concerns, the parties do not wish to expose their data to each other. Thus, distributed learning schemes are required to make use of the data of each participating party to collaboratively train an ML model. The datasets held by different parties may have different attributes, resulting in the so-called vertical partition of training data. That is to say, privacy-motivated DML is often applied in the scenarios with vertically partitioned datasets, with subsets of training data with different attributes held by different parties.

3.1.2   DML PLATFORMS

Because of the distributed and parallel computing architecture of DML, specialized ML platforms are required in order to reap the benefits of DML. There exist numerous commercial and open-source DML platforms. We introduce here some of the representative frameworks.

One of the most widely used distributed data processing systems for ML is Apache Spark MLlib [Apache MLlib, 2019]. MLlib is Apache Spark's scalable ML library. It is a memory-based DML framework and makes practical ML systems scalable and easy to deploy. MLlib offers distributed implementations of the conventional ML algorithms (as compared to DL), such as classification, regression, and clustering, etc. Apache DeepSpark offers implementation of distributed training framework for DL [DeepSpark, 2019].

Graph-based parallel processing is a relatively new approach for DML, which is also called graph parallelism in the context of DML (see Section 3.2.2). The platform GraphLab [Turi-Create, 2019, Low et al., 2010] offers scalable ML toolkits and implements fundamental algorithms like SGD and gradient descent with superior performance. Another graph parallelism-based computation platform is the Apache Spark GraphX, a new component in Spark, which implements a Pregel-like bulk-synchronous message passing application programming interface (API) [Apache GraphX, 2019], and Pregel is the parallel graph processing libraries from Google that is based on the Bulk Synchronous Processing (BSP) model [Malewicz et al., 2010].

The Distributed ML Toolkit (DMTK) released by Microsoft contains both algorithmic and system innovations [DMTK, 2019]. DMTK supports a unified interface for data parallelization, a hybrid data structure for big model storage, model scheduling for big model training, and automatic pipelining for high training efficiency.

DL requires training deep neural networks (DNNs) with massive number of parameters on a huge amount of data. Distributed and parallel computing is a perfect tool to take full advantage of the modern hardware. As for distributed DL, in addition to Apache DeepSpark, the popular DL frameworks, such as TensorFlow and PyTorch, all support distributed training and deployment.

TensorFlow supports distributed training of DNNs via tf.distribute, e.g., (i) allowing portions of the graph to be computed on different processes or even on different servers, and (ii) employing multiple processors or even servers to train the same model over different slices of input datasets [Distributed TensorFlow, 2019]. TensorFlow offers the possibility to split big models over many devices, carrying out the training in parallel on different devices if the models are too large to fit in the memory of a single device. In addition, this can be used to distribute computation to servers with powerful GPUs, and have other computations done on servers with more memory. With distributed TensorFlow, we can scale up distributed model training to hundreds of GPUs. We can massively reduce the experimentation (e.g., hyper-parameter tuning) time by running many experiments in parallel on many GPUs and servers.

The distributed package included in PyTorch (i.e., torch.distributed) enables researchers and practitioners to easily parallelize their computations across processes and clusters of machines [Arnold, 2019]. Similar to TensorFlow, distributed PyTorch allows a model to be logically split into several parts (i.e., some layers in one part and some in others), then placing them on different computing devices. PyTorch leverages the message passing semantics and allows each process to communicate data to any of the other processes. As opposed to the multiprocessing (e.g., torch.multiprocessing) package, processes in PyTorch can use different communication backends and are not restricted to being executed on the same machine.

3.2   SCALABILITY-MOTIVATED DML

In this section, we provide a brief review of the existing works on scalability-motivated DML methods. Readers are referred to Feunteun [2019], Ben-Nun and Hoefler [2018], Galakatos et al. [2018] and Bekkerman et al. [2012] for comprehensive surveys of DML schemes and the references therein for more technical details.

3.2.1   LARGE-SCALE MACHINE LEARNING

With the emergence of widespread communication and sensing devices, such as smartphones, portable gadgets, IoT sensors, and wireless cameras, data are ubiquitously available in enormous volumes. In this big data era, the bottleneck of ML methods has shifted from being able to infer from small training samples to dealing with large-scale high-dimensional datasets. With this trend shift, the ML community is faced with the challenge that the computation power and time do not scale well with the dataset size, making it impossible to learn from large-scale training samples with reasonable computation effort and time. We summarize in the following the major challenges that conventional ML methods are faced with when dealing with large-scale datasets.

1. **Memory shortage.** Conventional ML methods operate with the training samples entirely in one main memory. Therefore, if the computational complexity of the training samples exceed the main memory, the following problems may arise: (i) the trained model may not converge or may result in poor performance (such as bad precision or recall), and (ii) in the worst-case scenario, the ML models cannot be trained due to memory shortage.

2. **Unreasonable training time.** Some optimization process in ML algorithms may not scale well with respect to the training samples, such as Gaussian Mixture Model (GMM) and polynomial regression. As a result, when dealing with large-scale training samples, the time consumed by the training process may be too long for practical use. On top of training,

tuning hyper-parameters of ML models also takes a lot of time as we may need to try many different settings. Hence, if the training process takes too long, hyper-parameters tuning cannot be performed effectively, which may result in poor ML models.

Distributed ML algorithms are part of large-scale learning algorithms which has received considerable attention over the last few years, thanks to its ability to distribute the learning process onto several devices to scale up learning algorithms. Recent advances on DML make ML tasks on big data feasible, scalable, flexible, and more efficient.

3.2.2   SCALABILITY-ORIENTED DML SCHEMES

Excessive research efforts have been cast on presenting effective frameworks and methods for dealing with large-scale datasets and ML models. Particularly, training large-scale DL models is very time-consuming, with the training period ranging from days to even weeks. More recently, numerous research works have been carried out to push the frontiers of DML, aiming to reduce training time and cope with large-scale DL models. We review here some of the popular scalability-oriented DML schemes, covering data parallelism, model parallelism, graph parallelism, task parallelism, hybrid parallelism, and mixed parallelism.

Data Parallelism

The first instinct around DML is partitioning the training data into subsets, which are put on multiple computing entities that train the same model in parallel. This is known as the data parallelism approach, also known as the data-centric approach [Jia et al., 2019, Das, 2019, Wang, 2016]. In other words, data parallelism refers to processing multiple pieces (technically called shards) of training data through multiple replicas of the same model with different computing devices and communicating model information periodically. This approach can naturally scale up well with increasing amounts of training data. However, as a replica of the model (e.g., an entire DNN) has to reside on a single device, it cannot deal with ML models with high memory footprints.

There are mainly two common approaches for data parallelism-based distributed training, namely synchronous training and asynchronous training. With synchronous training, all computing entities train on replicas of the same model over different slices of the training data in synchronization, and the gradients (or model weights) produced by the computing entities are aggregated after each training step carried out by the entities. With asynchronous training, all entities independently train replicas of the same model over subsets of the training data and update model weights or gradients asynchronously. Typically, synchronous training is supported by the AllReduce architecture [Apache MapReduce, 2019, Fukuda, 2019], and asynchronous training by the parameter server architecture [Li et al., 2014].

Data parallelism can be used in the case that the training data is too large to be stored in a single device or to achieve faster training with parallel computing. Much work has been conducted for training DL models with distributed data. For example, the distributed frameworks, including DistBelief (which was later integrated into TensorFlow) from Google [Dean et al., 2012] and Project Adams [Chilimbi et al., 2014] from Microsoft, tend to train large-scale models with thousands of processors by utilizing both data and model parallelism.

Model Parallelism

As DL models are getting larger and larger, e.g., the BERT model [Devlin et al., 2019], we may face the problem that a DNN model cannot be loaded into the memory of a single computing entity. In such scenarios, we need to split the model and put parts of the model into different entities. This is called the model parallelism approach, also known as the model-centric approach [Jia et al., 2019, Das, 2019, Wang, 2016]. In other words, model parallelism refers to the case that a model (e.g., a DNN model) is being logically split into several parts (i.e., some layers in one part and some layers in other parts for a DNN model), then placing them in different computing devices. Although doing so does reduce execution time (asynchronous processing of data), it is usually employed to address memory constraints. Models with a very large number of parameters, which are difficult to fit into a single system due to high memory footprint, benefit from this type of strategy. For example, a single layer of a large DNN model can be fit into the memory of a single device and forward and backward propagation involves communication of output from one device to another in a serial fashion. We usually resort to model parallelism only if the model cannot fit into a single machine, not primarily to speed up the training process.

Existing works on model parallelism-based distributed training are extensive. One outstanding example is AMPNet, studied in Gaunt et al. [2018]. AMPNet was implemented on multi-core CPUs and it was shown that AMPnet training converged to the same accuracy as conventional synchronous training algorithms in a similar number of epochs, but took significantly shorter overall training time. A more recent example is OptCNN [Jia et al., 2018], which uses *layer-wise parallelism* for parallelizing convolutional neural networks (CNNs) with linear computation graphs. OptCNN allows each layer in a CNN to use an individual parallelization strategy. It was shown in Jia et al. [2018] that layer-wise parallelism outperforms state-of-the-art approaches by increasing training throughput, reducing communication costs, achieving better scalability to multiple GPUs, while maintaining original model accuracy.

Earlier exemplary works on model parallelism include Dean et al. [2012] and Kim et al. [2016], and Jia et al. [2019]. Particularly, in Dean et al. [2012], Google presented downpour SGD, which provides an asynchronous and distributed implementation of SGD. Downpour SGD combines data parallelism and model parallelism, which divides training samples among different machines, and each machine has a unique copy of the whole/partial network. DeepSpark was first proposed in Kim et al. [2016]. It allows distributed execution of both Caffe and Google's TensorFlow DL jobs on an Apache Spark cluster of machines [DeepSpark, 2019]. DeepSpark makes deploying large-scale parallel and distributed DL easy and intuitive for practitioners.

Graph Parallelism

As graph-based ML algorithms are fast-growing [Zhang et al., 2018], graph parallelism based DML approaches are also receiving more attention. Graph parallelism, also known as the graph-centric approach, is a relatively new technique to partition and distribute training data and execute ML algorithms that is orders of magnitude faster than data parallelism-based approaches [Tian et al., 2018, Wang, 2016, Low et al., 2010].

GraphLab, first studied in Low et al. [2010] as an improvement upon abstractions like MapReduce, implements asynchronous iterative algorithms with sparse computational dependencies while ensuring data consistency. It achieves a high degree of parallel performance. GraphLab is able to achieve excellent parallel performance on large scale real-world ML tasks.

More recently, Xiao et al. [2017] proposed a new distributed graph engine called TUX^2. TUX^2 is intentionally optimized for DML to support heterogeneity, with a Stale Synchronous Parallel model, and a new MEGA (Mini-batch, Exchange, GlobalSync, and Apply) model. TUX^2 puts forward the convergence of graph computation and DML, with a flexible graph model to express ML algorithms efficiently. Advances in graph computation and DML will allow more ML algorithms and optimization to be expressed and implemented easily and efficiently at scale.

Task Parallelism

Task parallelism, also known as the task-centric approach, covers the execution of computer programs across multiple processors on the same or multiple machines. It focuses on executing different operations in parallel to fully utilize the available computing resources in the form of processors and memory. One example of task parallelism would be an application of creating threads for doing parallel processing where each thread is responsible for performing a different operation. Examples of the big data frameworks that utilize task parallelism are Apache Storm [Apache Storm, 2019] and Apache YARN [Apache YARN, 2019].

It is common to combine task parallelism and data parallelism for DML. One outstanding example is Boehm et al. [2016], which presented a systematic approach for combining task parallelism and data parallelism for large-scale ML on top of MapReduce [Apache MapReduce, 2019]. The proposed framework enables users to specify task- and data-parallel ML algorithms in an easy and flexible way via a high-level primitive. The combined task and data parallelism on top of MapReduce opens ways to share cluster resources with other MapReduce based systems since the MapReduce scheduler provides global scheduling.

Hybrid Parallelism and Mixed Parallelism

In practical implementations of DML systems, we often need to combine different types of parallelism methods, resulting in the so-called hybrid parallelism, such as Apache YARN [Apache YARN, 2019] and SystemML [Pansare et al., 2018, Boehm et al., 2016] for both data and task parallelism. In fact, it is very common in practice to use both data and model parallelism simultaneously, such as Google downpour SGD [Dean et al., 2012] and the distributed DL framework proposed in Shrivastava et al. [2017]. Wang et al. [2018] to unify data, model, and hybrid parallelism via tensor tiling. The SOYBEAN system proposed in Wang et al. [2018] is a hybrid between data parallelism and model parallelism, and it performs automatic parallelization.

Under the broader umbrella of hybrid parallelism, there is also mixed parallelism, such as the work of Krizhevsky [2014] and Song et al. [2019]. This kind of parallelism is sometimes adopted for training large-scale DNNs, by distributing some layers using data parallelism and other layers using model parallelism. Readers are referred to Wang et al. [2018] and Song et al. [2019] for more information and related works on hybrid parallelism and mixed parallelism.

3.3 PRIVACY-MOTIVATED DML

The DML system can not only accelerate the computing of large-scale data, but also integrate data from different sites. In many practical areas, data are distributed to different clients, entities, and institutions. To collect more data to improve the performance, companies will also collect and analyze data from individuals, which comes with issues of user privacy and data security. For example, in medical applications, a hospital or medical institution is forbidden to share medical data according to regulations (e.g., HIPAA). Another example is that smart wearable devices are always collecting sensitive individual data, which are critical for wearable applications. However, sharing these data for model training also raises concerns about privacy leakage.

In a nutshell, sharing data and distributed computation is a trend in the era of big data, as it can (1) improve the computational efficiency, and (2) improve the model performance. In the meanwhile, the increasing awareness of privacy and data security requires a DML system to take privacy-preserving into consideration. Therefore, building a privacy-motivated DML system has become an important research direction. In this section, we start from a privacy-preserving decision tree example and further introduce several privacy-preserving techniques and their applications in a DML system.

For a privacy-preserving DML system, it generally protects some or all of the following information [Vepakomma et al., 2018].

1. Input training data;

2. Output predicted labels;

3. Model information, including parameters, architecture, and loss function; and

4. Identifiable information, such as which site a record comes from.

3.3.1 PRIVACY-PRESERVING DECISION TREES

The decision tree is an important kind of supervised ML algorithm, which is widely used in classification and regression. The learned model of a decision tree is explainable and understandable

to people. There are variants of decision trees, and ID3 [Quinlan, 1986] is one of the most famous of them. In distributed decision tree algorithms, it is normally divided into two categories
according to the data distribution, formally defined as follows.

1. **Horizontally** partitioned datasets:

$$\mathcal{X}_i = \mathcal{X}_j, \; \mathcal{I}_i \neq \mathcal{I}_j \;\; \forall \mathcal{D}_i \neq \mathcal{D}_j.$$

2. **Vertically** partitioned datasets:

$$\mathcal{X}_i \neq \mathcal{X}_j, \; \mathcal{I}_i = \mathcal{I}_j \;\; \forall \mathcal{D}_i \neq \mathcal{D}_j,$$

where $\mathcal{X}$ is the feature space of the data, and $\mathcal{I}$ is the sample space (i.e., the identification of
each sample) of the data. We next provide an explanation on this definition.

In the scenario of the horizontally partitioned dataset, each participant (noted as an entity) in the DML system owns different samples, and samples in all entities have the same
attribute categories. For example, the data collected by different wearable devices have the same
set of attributes since the sensors of the devices are the same. However, as a result of different
environments, the data samples collected by different entities should be different.

In the scenario of the vertically partitioned dataset, the attribute sets of data owned by
different entities are different, but these samples are possibly referring to the same group of
users. For example, the medical records of the same patient in different medical institutions
record different physiological indices or disease examination results.

For horizontally partitioned DML, the aggregation of samples is equivalent to enlarging
the dataset, while in the vertically partitioned scenario, it is similar to augmenting the features
of the samples. In either way, distributed training provides a manner to expand the dataset.

Different from other ML algorithms, the data partition is critical for decision trees, because the learning of a decision tree needs to determine the split of the attribute set, depending
on both the attribute category and the number of samples under a particular attribute with a
class label.

Lindell and Pinkas [2002] first propose a privacy-preserving distributed decision tree algorithm based on the horizontally partitioned dataset. They introduced an oblivious secure protocol which computes $(v_1 + v_2) \log(v_1 + v_2)$ without revealing each value to other participants.
This secure computation allows the distributed decision tree to privately calculate the node split
across the samples in different participants. Wang et al. [2006] and Du and Zhan [2002] first
addressed the problem of designing a vertically partitioned distributed decision tree with privacy
protection. However, their solutions assume all of the participants own the class attribute.

Fang and Yang [2008] completed their work by allowing only one entity in the vertically
partitioned distributed system to have the class attribute. Their work is based on ID3-tree, where
the learning of the tree is decomposed into different components including attribute checking,
distribution counts, class checking, attribute information gain checking, and information gain

computation. Each part of distributed computation is protected by secure protocols. Besides, this work provides a *loosely secure* version and a *completely secure* version, providing a trade-off between efficiency and security.

Cheng et al. [2019] proposed a vertically distributed boosting tree based on secure set intersection protocol and partially homomorphic encryption. They prove that their method is not only secure and privacy-preserving but also lossless. There are also privacy-preserving distributed decision trees using differential privacy (DP) to protect individual privacy by adding noise to the statistics [Jagannathan et al., 2009].

The development of privacy-motivated decision tree algorithms considers the data partition and utility of privacy-preserving tools. As a preliminary of privacy-preserving distributed ML systems, we briefly introduce some commonly used tools for privacy and security protections in the next subsection.

3.3.2   PRIVACY-PRESERVING TECHNIQUES

In privacy-motivated DML system, the popular tools to protect data privacy can be roughly categorized into the following two categories.

1. Obfuscation: to randomize or modify the data to conceal a certain level of privacy, e.g., DP.

2. Cryptographic Methods: to secure the distributed computation process without revealing input values to other participants, e.g., secure multi-party computation (MPC), including oblivious transfer (OT), secret sharing (SS), garbled circuit (GC), and homomorphic encryption (HE).

Please refer to Section 2.4 for a quick review of the aforementioned privacy-preserving techniques.

3.3.3   PRIVACY-PRESERVING DML SCHEMES

In the following, we give a quick review of representative works on privacy-preserving DML, emphasizing on how they utilize privacy protection tools mentioned above to protect the data and model security in a distributed environment. According to the aforementioned tools, we first summarize the DML algorithms using obfuscation and then introduce those algorithms that use cryptographic methods.

Chaudhuri and Monteleoni [2009] proposed a privacy-preserving logistic regression algorithm based on DP. They tackle the optimization over randomized data, making it possible to take a balance between model performance and privacy protection and make the privacy bound tighter. Following the definition given by Dwork [2008], they prove that their work guarantees ε-differential privacy, and provide a novel algorithm with better performance. In the proposed work, a randomized vector is generated using a Gamma function, which participates in the optimization of the logistic regression parameter θ. Moreover, they concluded that their work reveals the relation between perturbation-based privacy protection and regularization.

Wild and Mangasarian [2007] and Mangasarian et al. [2008] studied privacy-preserving support vector machines (PPSVMs) on horizontally and vertically partitioned datasets, respectively. They concealed the originally learned kernel with a *randomly generated kernel*, achieving comparable performance to the non-private SVMs. The privacy proof is based on the fact that there are infinite possible input data that can be recovered from the perturbed kernel. Therefore, sharing the perturbed kernel will not cause privacy leakage. However, these methods require participants to share the *randomly generated kernel*, limiting the application of these methods.

Apart from logistic regression, SVMs, and decision trees, perturbation-based privacy protection methods are also widely used in DL systems. With reference to the survey [Zhang et al., 2018], we selectively introduce some representative works. Song et al. [2013] proposed a differentially private stochastic gradient descent algorithm (DP-SGD), which clips the gradients and injects noise to them during training, so that the learned DL model preserves (ε, δ)-differential privacy. Different from previous works, Song et al. [2013] and Shokri and Shmatikov [2015] utilize another obfuscation method, i.e., a distributed selective stochastic gradient descent algorithm. It allows the local model to selectively share part of the parameters, avoiding information leakage and also preserving the performance of the joint learning model. Except for jointly learning a prediction model, Dwork [2011] proposed a differentially private autoencoder to learn the representation of local data.

DP is also used for unsupervised learning. Park et al. [2016] proposed a differentially private EM algorithm (DP-EM) based on moment perturbation. They utilize moment accountants [Abadi et al., 2016] and zCDP to reduce the magnitude of noise added to the EM process while maintaining the same level of privacy protection compared to the original analysis technique. In their work, they compare different randomization mechanisms and their composition settings, and find that in DP-EM, using Gaussian mechanism in every stage of EM can achieve the tightest privacy budget.

In conclusion, owing to the computational efficiency and implementation convenience, obfuscation-based privacy-preserving techniques are popular in privacy-motivated DML systems. Meanwhile, perturbation affects the utility of the data and model. In practice, researchers have to make a tradeoff between privacy protection and performance. Compared to obfuscation-based methods, cryptographic methods do not need to sacrifice data accuracy and model performance.

In Aono et al. [2016], authors utilize HE to protect the data during the training of logistic regression. Their method uses a two-degree approximation to the log-linear objective function, making the training process compatible with the additive HE method, which improves computational efficiency while maintaining a comparable performance. Besides, they claim that their output is compatible with DP. They also analyze the storage and computation complexity of their system, showing that their system supports large-scale distributed computation. Fienberg et al. [2006] also considered linear regression (LR) on the horizontally partitioned dataset, utilizing MPC method to aggregate the calculation. However, in their setting, the features are categorical, which means the computation space is small. Slavkovic et al. [2007] made significant progress, using *secure summation protocol* and *secure matrix multiplication* for the aggregation of distributed learning of LR, which supports both vertical and horizontal data partition.

Vaidya and Clifton [2004] designed a secure parameter sharing mechanism for privacy-preserving naive Bayes classifiers on vertically partitioned data, where each participant contributes to a conditionally independent probability. Each individual parameter is indistinguishable from random noise, while only the aggregation is meaningful. However, the extra computational complexity for secure computation is also significant. Yu et al. [2006] and Zhan and Matwin [2007] introduce *secure dot product* and *secure summation protocol* to protect the data during kernel computation in SVMs. Xu et al. [2015] embed the *secure summation protocol* into the Reducer of Hadoop system, implementing an efficient distributed SVM system that supports large-scale data. Similarly, Lin et al. [2005] use *secure summation protocol* to aggregate the local computation results for distributed EM algorithm, preventing data leakage from local computation results.

As for DL, one of the most representative work is SecureAggregation, proposed by Bonawitz et al. [2016]. This work is based on the federated learning algorithm, FedAvg, first adopted for federated learning by Google [McMahan et al., 2016b], further introducing secret sharing, and oblivious transfer to FedAvg, in a complex mobile environment where the communication is expensive, and the dropout and join-in of clients are frequent. To ensure data security, each leaving data is randomized, and only the aggregation of these shares is meaningful. To reduce the communication cost, they only exchange random seed using *secure key-exchange protocol* instead of random noise. To handle the challenge that clients can dropout unexpectedly, they use secret sharing so even some client are lost, the system can still recover the data using the remaining shares. Besides SecureAggregation, there are other DL algorithms that use MPC methods [Liu et al., 2016, Shokri and Shmatikov, 2015].

In addition, Mohassel and Zhang [2017] propose a set of privacy-preserving ML algorithm based on MPC methods, supporting LR, logistic regression, and SGD, and provide C++ implementation. Different from obfuscation-based privacy-preserving DML algorithms, cryptographic method-based methods emphasize on taking balance between computational and communication complexity, and security.

In the above, we have briefly introduced some representative privacy-motivated DML algorithms, as well as some widely used privacy-protection tools. For a more detailed treatment of this topic, the following surveys Vepakomma et al. [2018], Zhang et al. [2018] and Mendes and Vilela [2017] are recommended.

3.4   PRIVACY-PRESERVING GRADIENT DESCENT

Gradient descent method is one of the central algorithms in ML. Privacy-preserving gradient descent methods have been widely studied. In this section, we review different privacy-preserving techniques proposed for gradient descent. There is a trade-off between efficiency, accuracy and privacy protection. Developing good privacy-preserving gradient descent methods needs an artful balance of the efficiency-accuracy-privacy trade-off.

Methods preferring higher efficiency while facing less privacy concern may sacrifice data privacy for higher computational efficiency. For example, the gradients are sent to a coordinator in plain-text for model update in the gradient averaging approach [McMahan et al., 2016b], to trade privacy for efficiency without degrading the global learning accuracy. Methods aiming for the highest data privacy and security generally choose to use HE and MPC, which in turn leads to high computation complexity and communication overhead.

In addition to the approaches discussed in Chapter 2, there also exist some other privacy-preserving approaches with different privacy guarantees. In some cases, the privacy model only aims to guarantee that the raw form of input data of each party could not be revealed by the adversary. By providing such weak privacy guarantees, researchers aim to trade privacy for efficiency. Proposed approaches vary significantly.

Typical privacy-preserving gradient descent approaches include naive federated learning, algebraic approaches, sparse gradient update approaches, obfuscation approaches and cryptographic approaches (e.g., HE and MPC). Obfuscation approaches are based on randomization, generalization or suppression mechanisms (e.g., gradient quantization, DP, k-anonymity). Generally in the naive federated learning, algebraic approaches and sparse gradient update approaches, each party sends clear-text gradient to the coordinator for model update, which only protects raw data and yields weak privacy guarantee with quite high efficiency. The sparse gradient update approaches also trade accuracy for efficiency and privacy by updating a subset of entries in the gradient. Approaches based on randomization mechanism such as DP and Gaussian Random Projection (GRP) trade accuracy for privacy by adding random noise to the data or gradient. Approaches based on the generalization and suppression also trade accuracy for privacy by generalizing attributes or removing some instances. We review here some approaches for privacy-preserving gradient descent with roughly increasing privacy guarantees.

3.4.1   VANILLA FEDERATED LEARNING

Federated Averaging (FedAvg) was first employed for federated learning over horizontally partitioned dataset. In federated averaging, each party uploads clear-text gradient to a coordinator (or a trusted dealer, or a parameter server) independently, then the coordinator computes the average of the gradients and update the model. Finally, the coordinator sends the clear-text updated model back to each party [McMahan et al., 2016b]. When the dataset is vertically partitioned, the model is distributed among the parties. In gradient descent methods, the objective function can be decomposed into a differentiable function and a linearly separable function [Wan, 2007].

To conduct gradient descent, each party applies its data on its partial model to get intermediate computation results and send them to the coordinator in clear-text. The coordinator accumulates the intermediate results and evaluates the differentiable function to compute the loss and gradient. Finally, the coordinator updates the whole and send the updated partial model to each corresponding party. It is assumed that the coordinator is honest and incurious, and does not collude with any party. If the coordinator is corrupted, the gradient information of each party can be disclosed. Although the raw form of training data is not likely to be inferred from the gradient of each party, it has been demonstrated that it is possible to infer considerable information from the gradient uploaded from each party [Aono et al., 2018].

3.4.2   PRIVACY-PRESERVING METHODS

Algebraic Approaches

Algebraic approaches aim to protect raw training data by leveraging the algebraic properties of data transmitted. They preserve privacy by guaranteeing that there exist infinite valid input-output pairs of each honest party against the input and output of the adversaries, that is, the raw form of input data is protected. Wan [2007] proposed a secure gradient descent approach for two-party vertical federated learning by decomposing the target function into a differentiable function and a linearly separable function. In this approach, the model parameters of the two parties are mutually masked, and only the clear-text gradient is disclosed and used for model update. Such approaches implicitly radically assumes that each party has no knowledge of the records of the other party. If a small subset of records are disclosed to the other party (e.g., by data poisoning), the model can be easily disclosed via solving-equation attack.

To defend against the equation-solving attack, a secure two-party computation approach for vertical federated linear regression and classification is proposed by introducing the concept of *k-secure* [Du et al., 2004]. In this approach, the instances are aligned and the dependent attribute is in public. Arithmetic secret sharing is used here. As the addition operation in arithmetic secret sharing is done locally, thus it is information-theoretical secure. The two-party multiplication is conducted as follows. First, the two parties jointly generate a random invertible matrix $\mathbf{M}$. Then, one party A does matrix multiplication with its input matrix $\mathbf{A}$ to the left-side and right-side sub-matrix of $\mathbf{M}$ in turn, and sends the first $\mathbf{A}_1$ result to party B. The other party B does matrix multiplication with its input matrix to the top-side and bottom-side sub-matrix of $\mathbf{M}$ in turn, and send the second result $\mathbf{B}_2$ to party A. Finally, both parties compute $\mathbf{V}_a = \mathbf{A}_1 \cdot \mathbf{B}_1$ and $\mathbf{V}_b = \mathbf{A}_2 \cdot \mathbf{B}_2$, where $\mathbf{V}_a + \mathbf{V}_b = \mathbf{A} \cdot \mathbf{B}$.

The security of this protocol is based on the algebraic property that $2n^2$ equations cannot determine $n \times N$ variables, where $N >> n$. Although the gradient descent approach is not demonstrated in this study, it is straightforward to implement it according to what is discussed in Chapter 2.

Sparse Gradient Update Approaches

The sparse gradient update approaches preserve privacy by updating only a subset of gradients. Such approaches trade accuracy for efficiency and weak privacy. As the gradient is in the clear, they also trade privacy for efficiency. For example, Shokri and Shmatikov [2015] disclose only a subset of clear-text gradient parameters to the coordinator for model update. Strategies for improving communication efficiency include structured update and sketched update [Konecný et al., 2016a]. The structured update strategy only updates a sparse or a low-rank gradient matrix, and the sketched update strategy utilizes subsampling and quantization to eliminate the volume of gradients.

Multi-objective evolutionary computation is also explored in federated learning to learn sparse model parameters [Zhu and Jin, 2018]. Sparse gradient update and gradient compression are widely studied for stronger privacy preservation. However, there is little formal analysis of the privacy preserved by the gradient compression.

Obfuscation Approaches

Obfuscation approaches obfuscate data via randomization, generalization and suppression, which leads to an improvement of privacy at the cost of a deterioration of accuracy. In federated learning, local differential privacy (LDP) can be used by applying an additive noise mask to the gradient of each party. Jiang et al. [2019] propose an approach that applies independent Gaussian random projection (GRP) to protect the raw form of training data. Each participant first generates a Gaussian random matrix and project its raw training data for obfuscation. Then the obfuscated data are sent to the coordinator for model training. The privacy protected is only the raw form of the training data of each participant. The GRP used here also faces a problem of scalability in terms of both number of participants and attribute dimension. The gradient quantization strategy quantizes each gradient value to an adjacent value, which trades model accuracy for efficiency and privacy [Konecný et al., 2016a].

Cryptographic Approaches

The approaches discussed above disclose clear-text gradient of each party to the coordinator or each other parties. In contrast, cryptographic approaches leverage HE and MPC to preserve the privacy of the gradient of each party in the gradient descent procedure. The security models vary from security against honest-but-curious adversaries to security against malicious adversaries, and the corruption assumptions also vary a lot. In addition to the security model, the information disclosed by each approach also varies. Cryptographic approaches trade efficiency for privacy. As it could be very computation or communication inefficient, the approximations of nonlinear functions are generally introduced to further trade accuracy for efficiency.

The secure aggregation approaches introduce a coordinator that is only allowed to learn the clear-text average of a group of gradients. Secure aggregation preserves the privacy of each party's gradient via Shamir's threshold secret sharing scheme, so that the coordinator can only disclose the average of a group of gradients [Bonawitz et al., 2016]. However, when the coordinator and $n-1$ parties collude in a group with n parties, the input gradient can be easily disclosed. As such secure aggregation is anonymous, extreme poisoning attack may occur. Bonawitz et al. [2016] proposes "trimmed mean," where the gradients are trimmed coordinate-wisely to prevent extreme poisoning adversary.

Other cryptographic approaches introduce one or more non-colluding coordinators, while the coordinators are not allowed to learn anything about the gradient and the model. For HE-based approaches, this can be done by adding a random mask to the value to be decrypted [Liu et al., 2019]. For MPC-based approaches, the trusted dealer can be asked to generate computation-independent materials (e.g., Beaver triples) [Beaver, 1991]. When there are multiple non-colluding coordinators introduced, each party generates secret sharing of its private data and shares it with each coordinator accordingly [Mohassel and Zhang, 2017, Wagh et al., 2018]. The gradient descent is then conducted between the coordinators.

When it is infeasible to introduce a coordinator, some coordinator-free cryptographic approaches can be adopted, where the secure multi-party gradient descent is conducted so that each party can learn nothing beyond its output and its input. Existing MPC protocols secure against a majority of colluding malicious adversaries include SPDZ [Damård et al., 2011], SPDZ_{2k}, Overdrive [Keller et al., 2018], and MASCOT [Keller et al., 2016]. In such approaches, an offline phase is implemented where the Beaver triples are generated by MPC before secure multi-party gradient descent. Bonawitz et al. [2016] demonstrate an actively-secure MPC protocol based on SPDZ_{2k} for decision tree and SVM. The gradient descent functions can also be evaluated similarly based on the SPDZ_{2k} protocol.

3.5 SUMMARY

This chapter presents a brief introduction of the scalability-motivated DML and the privacy-motivated DML. The scalability-motivated DML is widely employed for addressing the computation resource and memory limitations in large-scale ML problems. Parallelism techniques (such as data parallelism, model parallelism, and hybrid parallelism) are the major choices for implementing and expanding the scalability-motivated DML systems. The privacy-motivated DML is primarily adopted for preserving user privacy and ensuring data security with decentralized data sources. MPC, HE and DP are the main privacy-preserving techniques for realizing the privacy-motivated DML. Privacy-preserving gradient descent methods have also been widely used for empowering the privacy-motivated DML.

DML has received abundant attention in past years, and has been fast-developed into open-source and commercial products. Yet, there are still practical challenges that the existing DML systems cannot address. Federated learning is a special type of DML, and it can further address the issues that the conventional DML systems are faced with and enable us to build privacy-preserving AI. We will elaborate with details in the subsequent chapters on federated learning.

C H A P T E R   4

Horizontal Federated Learning

In this chapter, we introduce horizontal federated learning (HFL), covering the concept, architecture, application examples, and related works, as well as open research challenges.

4.1   THE DEFINITION OF HFL

HFL, a.k.a. sample-partitioned federated learning, or example-partitioned federated learning [Kairouz et al., 2019], can be applied in scenarios in which datasets at different sites share overlapping feature space but differ in sample space, as illustrated in Figure 4.1. It resembles the situation that data is horizontally partitioned inside a tabular view. In fact, the word "horizontal" comes from the term "horizontal partition," which is widely used in the context of the traditional tabular view of a database (e.g., rows of a table are horizontally partitioned into different groups and each row contains complete data features). For example, two regional banks may have very different user groups from their respective regions, and the intersection set of their users is very small. However, their business models are very similar. Hence, the feature spaces of their datasets are the same. Formally, we summarize the conditions for HFL as:

$$\mathcal{X}_i = \mathcal{X}_j, \ \mathcal{Y}_i = \mathcal{Y}_j, \ I_i \neq I_j, \ \forall \mathcal{D}_i, \mathcal{D}_j, i \neq j, \tag{4.1}$$

where the data feature space and label space pair of the two parties, i.e., $(\mathcal{X}_i, \mathcal{Y}_i)$ and $(\mathcal{X}_j, \mathcal{Y}_j)$, are assumed to be the same, whereas the user identifiers I_i and I_j are assumed to be different; $\mathcal{D}_i$ and $\mathcal{D}_j$ denote the datasets of the ith party and the jth party, respectively.

Security of an HFL system. An HFL system typically assumes honest participants and security against an honest-but-curious server [Phong et al., 2018, Bonawitz et al., 2017]. That is, only the server can compromise the user privacy and data security of the participants.

Shokri and Shmatikov [2015] proposed a collaborative deep learning (DL) scheme where participants train models independently and share only subsets of model parameter updates, which is a special form of HFL. In 2016, researchers at Google proposed an HFL-based solution for Android smartphone model updates [McMahan et al., 2016a]. In this framework, a single Android smartphone updates the model parameters locally and uploads the model parameters to the Android cloud, thus jointly training the federated model together with other Android smartphones.

A secure aggregation scheme for protecting the privacy of the user model updates under this federated learning framework was introduced in Bonawitz et al. [2017]. More recently,

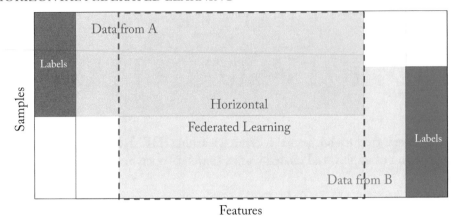

Figure 4.1: Illustration of HFL, a.k.a. sample-partitioned federated learning [Yang et al., 2019].

Phong et al. [2018] applied additively homomorphic encryption for model parameter aggregation to provide security against an untrustworthy central server.

In Smith et al. [2017], a multi-task style federated learning system is proposed to allow multiple sites to complete different tasks, while sharing knowledge and preserving security. Their proposed multi-task learning model can also address the issues of high communication costs, stragglers, and fault tolerance.

In McMahan et al. [2016a], the authors proposed a secure client-server structure where the federated learning system partitions data by users, and allows models built at client devices to collaborate at the server site to build a global federated model. The process of model building ensures that there is no data leakage. Likewise, in Konecný et al. [2016b], the authors proposed methods to reduce the communication cost to facilitate the training of federated models based on data distributed over mobile clients. More recently, a compression approach called Deep Gradient Compression [Lin et al., 2018] is proposed to greatly reduce the communication bandwidth in large-scale distributed model training.

Security proof has been provided in these works. Recently, another security model considering malicious user [Hitaj et al., 2017] is also proposed, posing additional privacy challenges. At the end of federated training, the aggregated model and the entire model parameters are exposed to all participants.

4.2   ARCHITECTURE OF HFL

In this section, we describe two popular architectures for HFL systems, namely the client-server architecture and the peer-to-peer architecture.

4.2.1 THE CLIENT-SERVER ARCHITECTURE

A typical client-server architecture of an HFL system is shown in Figure 4.2, which is also known as master-worker architecture. In this system, K participants (also known as clients or users or parties) with the same data structure collaboratively train a machine learning (ML) model with the help of a server (also known as parameter server or aggregation server or coordinator). A typical assumption is that the participants are honest whereas the server is honest-but-curious. Therefore, the aim is to prevent leakage of information from any participants to the server [Phong et al., 2018]. The training process of such an HFL system usually consists of the following four steps.

- **Step 1**: Participants locally compute training gradients, mask a selection of gradients with encryption [Phong et al., 2018], differential privacy [Abadi et al., 2016], or secret sharing [Bonawitz et al., 2017] techniques, and send the masked results to the server.

- **Step 2**: The server performs secure aggregation, e.g., via taking weighted average.

- **Step 3**: The server sends back the aggregated results to the participants.

- **Step 4**: The participants update their respective models with the decrypted gradients.

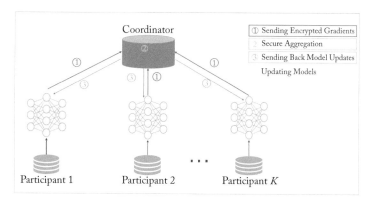

Figure 4.2: Exemplary client-server architecture for an HFL system [Yang et al., 2019].

Iterations through the above steps continue until the loss function converges or until the maximum number of allowable iterations is reached or until the maximum allowable training time is reached. This architecture is independent of specific ML algorithms (e.g., logistic regression and DNN, and all participants will share the same final model parameters.

Note that in the above steps, it is described that the participants send gradients to the server, which in turn, aggregates the received gradients. We call this approach *gradient averaging* [Tang et al., 2019, Su and Chen, 2018]. Gradient averaging is also known as synchronous stochastic gradient descent (SGD) or federated SGD (FedSGD) [McMahan et al., 2016a, Chen et al., 2017]. Alternatively, instead of gradients, the participants can share model weights. That is, participants locally compute model weights and send them to the server [Phong and Phuong, 2019]. The server aggregates the received local model weights and sends the aggregated results back to the participants. We call this approach *model averaging* [McMahan et al., 2016a, Yu et al., 2018, Xu et al., 2018]. In the extreme case, in which model parameters are averaged after each weight update carried out locally at the participants and the participants all start with the same initial model weights, model averaging is equivalent to gradient averaging [Su and Chen, 2018, McMahan et al., 2016a]. We summarize the comparison between gradient averaging and model averaging in Table 4.1. Note that both gradient averaging and model averaging are referred to as federated averaging (`FedAvg`) in McMahan et al. [2016a].

Table 4.1: Comparison between gradient averaging and model averaging [Tang et al., 2019, Su and Chen, 2018]

Method	Advantage	Disadvantage
Gradient averaging	Accurate gradient information Guaranteed convergence	Heavy communication Require reliable connection
Model averaging	Not bound to SGD Tolerance of update loss Infrequent synchronization	No guarantee of convergence Performance loss

The above architecture is able to prevent data leakage against a semi-honest server, if gradient aggregation is performed with secure multi-party computation [Bonawitz et al., 2017] or additively homomorphic encryption [Phong et al., 2018]. However, it may be vulnerable to attacks by a malicious participant training a Generative Adversarial Network (GAN) in the collaborative learning process [Hitaj et al., 2017].

The client-server architecture appears similar to the architecture of a distributed machine learning (DML) system, especially the *data-parallel* paradigm (see also Section 3.2) of DML. HFL also resembles geo-distributed machine learning (GDML) [Xu et al., 2018, Cano et al., 2016, Hsieh et al., 2017]. A parameter server [Li et al., 2014, Ho et al., 2013] is a typical element in DML. As a tool to accelerate the training process, the parameter server stores data on distributed working nodes, allocates data and computing resources through a central scheduling node, so as to train the model more efficiently. For HFL, a data owner is a working party. It has full autonomy to operate on its local data, and can decide when and how to join and contribute to an HFL system. In the parameter server paradigm [Li et al., 2014, Ho et al., 2013], the central node always takes control, while an HFL system is faced with a more complex learning environment. Further, HFL takes into account data privacy protection during model training. Effective measures to protect data privacy can better cope with the increasingly stringent user privacy and data security requirements coming up in the near future. Finally, in an HFL system, the data held of the different participants are not identically distributed in most of the practical applications, while in a DML system, the data held by the different computing nodes normally follow the same distribution.

4.2.2   THE PEER-TO-PEER ARCHITECTURE

In addition to the client-server architecture discussed above, an HFL system can also make use of the peer-to-peer architecture shown in Figure 4.3 [Zantedeschi et al., 2019, Chang et al., 2017, 2018, Phong and Phuong, 2019]. Under the peer-to-peer architecture, there is no central server or coordinator. In such scenarios, the K participants of an HFL system are also called trainers or distributed trainers or workers. Each trainer is responsible for training the same ML or DL model (e.g., a DNN model) using only its local data. Further, the trainers need secure channels to transfer the model weights to each other. To ensure secure communications between any two trainers, security measures, such as public key based encryption schemes, can be adopted.

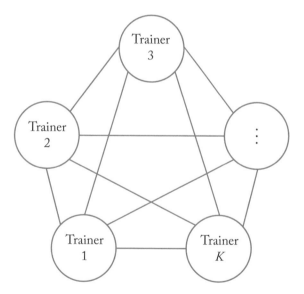

Figure 4.3: Exemplary peer-to-peer architecture for an HFL system.

Since there is no central server, the trainers must agree to the order of sending and receiving model weights in advance. There are mainly two ways to do this.

- **Cyclic transfer.** In the cyclic transfer mode, the trainers are organized into a chain. The first trainer (i.e., the top of the chain) sends its current model weights to its downstream trainer. One trainer receives model weights from its upstream trainer, and it updates the received model weights using mini-batches of training data from its own dataset. Then, it sends the updated model weights to its downstream trainer. For example, from trainer 1 to trainer 2, from trainer 2 to trainer 3, ..., from trainer $(K - 1)$ to trainer K, and from trainer K back to trainer 1. This procedure is repeated until the model weights converge or until the maximum number of iterations is reached or until the maximum allowable training time is reached.

- **Random transfer.** The kth trainer selects a number i from $\{1, \ldots, L\} \setminus \{k\}$ at random with equal probability, and sends its model weights to trainer i. When the ith trainer receives model weights from the kth trainer, it updates the received model weights using mini-batches of training data from its own dataset. Then, the ith trainer also selects a number j in $\{1, \ldots, L\} \setminus \{i\}$ at random and with equal probability, and sends its model weights to trainer j. This procedure takes place concurrently among the K trainers until the trainers agree that the model weights have converged or until the maximum allowable training time is reached. This method is also known as Gossip Learning [Hardy et al., 2018, Hegedüs et al., 2019].

Sharing model weights is used as an example in the above descriptions. It is also possible for the trainers to share gradients, such as using a gossip SGD-based approach, see, e.g., Liu et al. [2018] and Daily et al. [2018], for more details.

Compared with the client-server architecture, the obvious advantage of the peer-to-peer architecture is the possibility to remove the central server (also known as server, parameter server, aggregation server, or coordinator), which may not be available in practical applications, and it clears the chance of leaking information to the server. However, there are several disadvantages. For instance, in the cyclic transfer mode, since there is no central server, weight parameters are updated serially rather than in parallel batches, which takes more time to train a model.

4.2.3   GLOBAL MODEL EVALUATION

In HFL, model training and evaluation are carried out distributively at each participant, and it is impossible to access the datasets of the participants. As a consequence, each participant can easily test the mode performance using its local testing dataset to get the local model performance, but it takes more efforts to get the global model performance across all participants. Here, local model performance means the performance of an HFL model examined on the local test dataset of a single participant, and global model performance refers to the performance of an HFL model evaluated on the test datasets of all the participants. Model performance may be expressed in terms of accuracy, precision, recall, and area under the receiver operating characteristic curve (AUC), etc. For ease of elaboration, we use a two-class classification task as an example to explain how we can obtain the global model performance in HFL.

For the client-server architecture, the participants and the coordinator can cooperate to get the global model performance. During model training process and after model training being completed in HFL, we can obtain the global model performance according to the following steps.

- **Step 1:** The kth participant evaluates the performance of the current HFL model using its local test dataset. For the two-class classification task, this step generates the local model evaluation results such as $(N_{TP}^k, N_{FP}^k, N_{TN}^k, N_{FN}^k)$, where N_{TP}^k, N_{FP}^k, N_{TN}^k, and N_{FN}^k denote the number of true positive results, the number of false positive results, the number of true negative results, and the number of false negative results, respectively, for $k = 1, 2, \ldots, K$.

- **Step 2:** The kth participant sends the local model evaluation results $(N_{TP}^k, N_{FP}^k, N_{TN}^k, N_{FN}^k)$ to the coordinator, for $k = 1, 2, \ldots, K$.

- **Step 3:** After collecting the local model evaluation results of the K participants, i.e., $\{(N_{TP}^k, N_{FP}^k, N_{TN}^k, N_{FN}^k)\}_{k=1}^{K}$, the coordinator can calculate the global model performance. For example, for the two-class classification task, the global model recall can be computed as: $\frac{\sum_{k=1}^{K} N_{TP}^k}{\sum_{k=1}^{K}(N_{TP}^k + N_{FN}^k)}$.

- **Step 4:** The coordinator then sends the computed global model performance (e.g., accuracy, precision, recall, and AUC, etc.) back to all the participants.

For the peer-to-peer architecture, since there is no central coordinator, it would be more complicated to obtain global model performance. One possible way is to pick one of the trainers to serve as a temporary coordinator. Then, we can follow the above procedure proposed for the client-server architecture to obtain the global model performance for the peer-to-peer architecture. This method is recommended for evaluating the final HFL model after the training is completed. However, if we apply this method during the training process, it would overburden the temporary coordinator, which may not be acceptable if the trainers are mobile or IoT devices with limited resources (e.g., battery). One possible way to remedy this issue is to ask the trainers to take turns to act as the temporary coordinator.

4.3 THE FEDERATED AVERAGING ALGORITHM

In McMahan et al. [2016a,b], the federated averaging (FedAvg) algorithm was employed for federated model training in HFL systems. We review the FedAvg algorithm and its secured version in this section, assuming a client-server architecture. Note that the FedAvg algorithm is also known as parallel restarted SGD and local SGD [Yu et al., 2018, Haddadpour et al., 2019], as opposed to parallel mini-batch SGD.

4.3.1 FEDERATED OPTIMIZATION

The optimization problem arising from federated learning is referred to as federated optimization [Li et al., 2019, McMahan et al., 2016b], so to name it differently from distributed optimization. In fact, federated optimization has several key properties that differentiate it from a conventional distributed optimization problem [McMahan et al., 2016b, Xu et al., 2018, Cano et al., 2016].

- **Non-independent identical distributions (Non-IID) of datasets.** For distributed optimization within a data center, it is possible to ensure that different computing nodes have IID datasets so that all local updates look very similar. In federated optimization, this cannot be guaranteed. The data owned by different participants may follow completely different distributions, i.e., we cannot make IID assumptions about the decentralized datasets in federated learning [Li et al., 2019, Liu et al., 2018, Sattler et al., 2019]. For example, while similar participants might have similar local training data, two randomly picked participants might produce very different model weight updates or gradient updates.

- **Unbalanced number of data points.** For distributed optimization within a data center, it is possible to divide the data equally among the computing nodes. However, in realistic scenarios, different participants usually have very different volumes of training datasets [Chen et al., 2019, Li et al., 2018, Duan, 2019]. For example, some participants may hold only a handful of data points, while others might have large amounts of data.

- **Huge number of participants.** For distributed optimization within a data center, the number of parallel computing nodes can easily be controlled. However, since ML or DL generally requires a lot of data, the applications that use federated learning may need to involve many participants, especially with mobile devices [Bonawitz and Eichner et al., 2019]. Every one of these participants can theoretically participate in federated learning, making it far more distributed than that within a data center.

- **Slow and unreliable communication links.** In a data center, it is expected that nodes can communicate quickly with each other and that packets are almost never lost. However, in federated learning, communication between clients and the server relies on existing Internet connections. For example, uploads (from client to server) are typically going to be much slower than downloads, especially if the connection is from a mobile terminal. Some clients might also temporally lose connections to the Internet [Tang et al., 2019, Hartmann, 2019].

To address the above challenges faced in federated optimization, McMahan et al. first adopted the `FedAvg` algorithm for federated optimization [McMahan et al., 2016b]. The focus of `FedAvg` [McMahan et al., 2016a,b] is on non-convex objective functions commonly seen when training DNNs. `FedAvg` is applicable to any finite-sum objective function of the following form:

$$f(w) = \frac{1}{n} \sum_{i=1}^{n} f_i(w), \tag{4.2}$$

where n denotes the number of data points and $w \in R^d$ represents model parameters (e.g., model weights of a DNN) of dimension d.

For an ML or DL problem, we typically take $f_i(w) = \mathcal{L}(x_i, y_i; w)$, which is the loss of the prediction on sample (x_i, y_i) for the given model parameters w, where x_i and y_i denote the ith data point and the corresponding label, respectively.

Assume that there are K participants (also known as data owners or clients) in an HFL system, with $\mathcal{D}_k$ denoting the dataset owned by the kth participant, with $\mathcal{P}_k$ being the set of indexes of data points on client k. Define $n_k = |\mathcal{P}_k|$ as the cardinality of $\mathcal{P}_k$. That is, it is assumed that the ith participant has n_k training data points. As a result, considering there are K participants, Equation (4.2) can be rewritten as

$$f(w) = \sum_{k=1}^{K} \frac{n_k}{n} F_k(w) \quad \text{where} \quad F_k(w) = \frac{1}{n_k} \sum_{i \in \mathcal{P}_k} f_i(w). \tag{4.3}$$

When the data points owned by the K participants are independent and identically distributed (IID), then we have $\mathbb{E}_{\mathcal{D}_k}[F_k(w)] = f(w)$, where the expectation $\mathbb{E}_{\mathcal{D}_k}[\cdot]$ is taken over the set of data points owned by the kth participant. This IID assumption is typically made by distributed optimization algorithms in DML paradigm. If the IID assumption does not hold, which is known as the non-IID setting described above, the loss function $F_k(\cdot)$ maintained at the kth participant could be an arbitrarily bad approximation of the function $f(\cdot)$ [Goodfellow et al., 2016, Zhao et al., 2018, Sattler et al., 2019].

SGD and its variants (e.g., mini-batch gradient descent) are the most popular optimization algorithms for DL [Zhang et al., 2019]. Many advances on DL can be understood as adapting the structure of the model (and hence the loss function) to be more amenable to optimization by simple gradient-based methods [Goodfellow et al., 2016, Zhang et al., 2019]. In light of the widespread applications of DL, it is natural that we also develop new algorithms for federated optimization starting from SGD [McMahan et al., 2016b].

SGD can be applied naively to federated optimization, where a single mini-batch gradient calculation (e.g., on a randomly selected subset of participants) is performed during each round of federated training. Here, "one round" refers to the operations of sending updates from the participants to the server and from the server back to the participants, i.e., including the Steps 1–4 of Figure 4.2. This approach is computationally efficient, but requires very large number of communication rounds of training to produce satisfactory models, e.g., even using an advanced approach like batch normalization (BN) [Ioffe and Szegedy, 2015] training on the MNIST dataset requires 50,000 rounds on mini-batches of size 60 [McMahan et al., 2016b].

For DML, with parallel training within data centers, or computing-clusters, communication costs are relatively small, and computational costs dominate. Recent approaches focus on applying graphics processing units (GPUs) to lower these costs. In contrast, in federated learning, communication costs dominate as communication takes place over the Internet or wide area networks (WANs), even with wireless and mobile networks.

In federated learning, a single on-site dataset is usually small, compared to the total dataset size, and modern terminals (such as smartphones) have relatively fast processors, even including

GPUs. As a result, computation cost is negligible compared to communication costs for many model types in federated learning. Hence, we may use additional computation to decrease the number of rounds of communication needed to train a model. The following are two primary ways to add computation [McMahan et al., 2016b].

- **Increased parallelism.** We can engage more participants working independently in between client-server communication rounds.

- **Increased computation on each participant.** Rather than performing a simple computation like a gradient calculation, each client performs a more complex calculation in between communication rounds, such as performing multiple model weight update over a training epoch.

4.3.2   THE FEDAVG ALGORITHM

As articulated in McMahan et al. [2016b], the `FedAvg` algorithm family allows us to add computation using both approaches outlined above. The amount of computation is controlled by three key parameters, namely: (1) ρ, the fraction of clients that perform computation during each round; (2) S, the number of training steps each client performs over its local dataset during each round (i.e., the number of local epochs); and (3) M, the mini-batch size used for the client updates. We use $M = \infty$ to indicate that the full local dataset is treated as a single mini-batch.

We may set $M = \infty$ and $S = 1$ to produce a form of SGD with a varying mini-batch size. This algorithm selects a ρ-fraction of participants during each round, and computes the gradient and the loss function over all the data held by these participants. Therefore, in this algorithm, ρ controls the *global* batch size, with $\rho = 1$ corresponding to the full-batch gradient descent using all data held by all participants. Since we still select batches by using all the data on the chosen participants, we refer to this simple baseline algorithm as FederatedSGD. While the batch selection mechanism is different from selecting a batch by choosing individual examples uniformly at random, the batch gradients g computed by the FederatedSGD algorithm still satisfy $\mathbb{E}[g] = \nabla f(w)$, provided that the datasets held at different participants are IID.

It is commonly assumed that the coordinator or server has the initial ML model, and the participants know the settings of the optimizer. For a typical implementation of distributed gradient descent with a fixed learning rate η, in the tth round of global model weight update, the kth participant computes $g_k = \nabla F_k(w_t)$, the average gradient on its local data points at the current model weight w_t, and the coordinator aggregates these gradients and applies the update of model weights according to McMahan et al. [2016b]:

$$w_{t+1} \leftarrow w_t - \eta \sum_{k=1}^{K} \frac{n_k}{n} g_k, \qquad (4.4)$$

where $\sum_{k=1}^{K} \frac{n_k}{n} g_k = \nabla f(w_t)$, provided that the data points held at different participants are IID. The coordinator can then send the updated model weights w_{t+1} back to the participants.

Algorithm 4.1 The FedAvg Algorithm (adapted from McMahan et al. [2016b])

1: **The coordinator executes:**
2:    Initializes model weight w_0, and broadcasts the initial model weight w_0 to all participants.

3: **for** each global model update round $t = 1, 2, \ldots$ **do**
4:     The coordinator determines C_t, which is the set of randomly selected $\max(K\rho, 1)$ participants.
5:     **for** each participant $k \in C_t$ **in parallel do**
6:         Updates model weight locally: $w_{t+1}^k \leftarrow$ **Participant Update** $(k, \overline{w}_t)$.
7:         Sends the updated model weight w_{t+1}^k to the coordinator.
8:     **end for**
9:     The coordinator aggregates the received model weights, i.e., taking the weighted average of the received model weights: $\overline{w}_{t+1} \leftarrow \sum_{k=1}^{K} \frac{n_k}{n} w_{t+1}^k$
10:    The coordinator checks whether the model weights converges. If yes, then the coordinator signals the participants to stop.
11:    The coordinator broadcasts the aggregated model weights $\overline{w}_{t+1}$ to all the participants.
12: **end for**

13: **Participant Update** $(k, \overline{w}_t)$:
       (This is executed by participant k, $\forall k = 1, 2, ..., K$.)
14: Obtain the latest model weight from the server, i.e., set $w_{1,1}^k = \overline{w}_t$
15: **for** each local epoch i from 1 to number of epochs S **do**
16:    batches $\leftarrow$ randomly divides dataset $\mathcal{D}_k$ into batches of size M.
17:    Obtain the local model weight from last epoch, i.e., set $w_{1,i}^k = w_{B,i-1}^k$
18:    **for** batch index b from 1 to number of batches $B = \frac{n_k}{M}$ **do**
19:        Computes the batch gradient g_k^b.
20:        Updates model weights locally: $w_{b+1,i}^k \leftarrow w_{b,i}^k - \eta g_k^b$.
21:    **end for**
22: **end for**
23: Obtains the local model weight update $w_{t+1}^k = w_{B,S}^k$, and sends it to the coordinator.

Alternatively, the coordinator can send the averaged gradients $\overline{g}_t = \sum_{k=1}^{K} \frac{n_k}{n} g_k$ back to the participants, and the participants calculate the updated model weights w_{t+1} according to Equation (4.4). This method is called *gradient averaging* [Tang et al., 2019, Su and Chen, 2018].

It is straightforward to show that an equivalent approach is given by [McMahan et al., 2016b]

$$\forall k, \ w_{t+1}^k \leftarrow \overline{w}_t - \eta g_k \tag{4.5}$$

$$\overline{w}_{t+1} \leftarrow \sum_{k=1}^{K} \frac{n_k}{n} w_{t+1}^k. \tag{4.6}$$

That is, each client locally takes one step (or multiple steps) of gradient descent on the current model weights $\overline{w}_t$ using its local data according to Equation (4.5), and sends the locally updated model weights w_{t+1}^k to the server. The server then takes a weighted average of the resulting models according to Equation (4.6), and sends the aggregated model weights $\overline{w}_{t+1}$ back to the participants. This method is called *model averaging* [McMahan et al., 2016b, Yu et al., 2018].

The model averaging variant of the FedAvg algorithm is summarized in Algorithm 4.1. Once the algorithm is written in this way, it is natural to ask what happens when the participant iterates the local update (see Equation (4.5)) multiple times before going into the averaging step? For a participant with n_k local data points, the number of local updates per round is given by $u_k = \frac{n_k}{M} S$. The complete pseudo-code of the FedAvg algorithm with model averaging is given in Algorithm 4.1.

However, for general non-convex objectives, model averaging in the model weight space may produce an arbitrarily bad model, and it may not even converge at all [Tang et al., 2019, Su and Chen, 2018]. Luckily, recent work indicates that in practice, the loss surfaces of sufficiently over-parameterized DNNs are surprisingly well behaved, and in particular less prone to bad local minima than previously thought [Goodfellow et al., 2015]. When we start two models *from the same random initialization* and then, again train each independently on a different subset of the data (as described above), it has been found out empirically that model averaging based approach works surprisingly well [McMahan et al., 2016a,b, Yu et al., 2018]. The success of dropout training may provide some intuitions for the success of the federated model averaging scheme. Dropout training can be interpreted as averaging models of different architectures that share model weights, and the inference-time scaling of the model weights is analogous to the model averaging used in Srivastava et al. [2014].

4.3.3   THE SECURED FEDAVG ALGORITHM

The plain FedAvg algorithm in the form of Algorithm 4.1 exposes plaintext intermediate results, such as gradients from an optimization algorithm like SGD or model weights of DNNs, to the coordinator. No security guarantee is provided against the coordinator, and the leakage of gradients or model weights may actually leak important data and model information [Phong et al., 2018] if the data structure is also exposed. We can leverage privacy-preserving techniques,

---

Algorithm 4.2 The Secured FedAvg Algorithm (model averaging with AHE)

1: **The coordinator executes:**
2:     Initializes model weight w_0, and broadcasts the initial model weight w_0 to all participants.

3: **for** each global model update round $t = 1, 2, \ldots$ **do**
4:     The coordinator determines $\mathcal{C}_t$, which is the set of randomly selected $\max(K\rho, 1)$ participants.
5:     **for** each participant $k \in \mathcal{C}_t$ **in parallel do**
6:         Updates model weight locally: $[[w_{t+1}^k]] \leftarrow$ **Participant Update** $(k, [[\overline{w}_t]])$.
7:         Sends the updated model weight $[[w_{t+1}^k]]$ and the corresponding loss $\mathcal{L}_{t+1}^k$ to the coordinator.
8:     **end for**
9:     The coordinator aggregates the received model weights, i.e., taking the weighted average of the received model weights: $[[\overline{w}_{t+1}]] \leftarrow \sum_{k=1}^{K} \frac{n_k}{n} [[w_{t+1}^k]]$ (With slight abuse of notations for visual comfort, these are *operations over ciphertexts*, see, e.g., Paillier [1999].)

10:     The coordinator checks whether the loss $\sum_{k \in \mathcal{C}_t} \frac{n_k}{n} \mathcal{L}_{t+1}^k$ converges or the maximum number of rounds is reached. If yes, then the coordinator signals the participants to stop.
11:     The coordinator broadcasts the aggregated model weights $[[\overline{w}_{t+1}]]$ to all the participants.
12: **end for**

13: **Participant Update** $(k, [[\overline{w}_t]])$:
    (This is executed by participant k, $\forall k = 1, 2, ..., K$.)
14: Decrypts $[[\overline{w}_t]]$ to obtain $\overline{w}_t$.
15: Obtain the latest model weight from the server, i.e., set $w_{1,1}^k = \overline{w}_t$
16: **for** each local epoch i from 1 to number of epochs S **do**
17:     batches $\leftarrow$ randomly divides dataset $\mathcal{D}_k$ into batches of size M.
18:     Obtain the local model weight from last epoch, i.e., set $w_{1,i}^k = w_{B,i-1}^k$
19:     **for** batch index b from 1 to number of batches $B = \frac{n_k}{M}$ **do**
20:         Computes the batch gradient g_k^b.
21:         Updates model weights locally: $w_{b+1,i}^k \leftarrow w_{b,i}^k - \eta g_k^b$.
22:     **end for**
23: **end for**
24: Obtains the local model weight update $w_{t+1}^k = w_{B,S}^k$.
25: Performs AHE on w_{t+1}^k to get $[[w_{t+1}^k]]$, and sends $[[w_{t+1}^k]]$ and the corresponding loss $\mathcal{L}_{t+1}^k$ to the coordinator.

such as the widely used methods described in Chapter 2, to ensure user privacy and data security in `FedAvg`.

As an illustrative example, we use additively homomorphic encryption (AHE) [Acar et al., 2018] (e.g., the Paillier algorithm [Paillier, 1999] or the learning with errors- (LWE) based encryption [Phong et al., 2018]) to enhance the security feature of the `FedAvg` algorithm.

Recall that AHE is a semi-homomorphic encryption algorithm that only supports the addition and multiplication operations (i.e., additional homomorphism and multiplicative homomorphism [Paillier, 1999]). For ease of reference, we summarize here the key properties of AHE. Let $[[u]]$ and $[[v]]$ denote the homomorphic encryption of u and v, respectively. With AHE, the following holds (see Section 2.4.2):

- **Addition:** $Dec_{sk}([[u]] \oplus [[v]]) = Dec_{sk}([[u + v]])$, where "$\oplus$" may represent multiplication of the ciphertexts (see, e.g., Paillier [1999]).

- **Scalar multiplication:** $Dec_{sk}([[u]] \odot n) = Dec_{sk}([[u \cdot n]])$, where "$\odot$" may represent taking the power of n of the ciphertext (see, e.g., Paillier [1999]).

Thanks to these two nice properties of AHE, we can directly apply AHE to the `FedAvg` algorithm to ensure security against the coordinator/server.

Specifically, by comparing Algorithm 4.1 with Algorithm 4.2, it can be observed that security measures, such as AHE, can be easily added on top of the original `FedAvg` algorithm to provide secure federated learning. It was shown in Phong et al. [2018] that, under certain conditions, the secured `FedAvg` algorithm in Algorithm 4.2 leaks no information of the participants to an *honest-but-curious* coordinator, provided that the underlying homomorphic encryption scheme is chosen-plaintext attack (CPA) secure. In other words, Algorithm 4.2 ensures honest-but-curious security against the coordinator. With AHE, the data and the model itself are not transmitted in plaintext form. Hence, there is almost no possibility of leakage at the raw data level. However, encryption and decryption operations will increase the computational complexity, and transmission of cyphertext will introduce extra communication overhead. Another drawback of AHE is that polynomial approximations (e.g., using first-order Taylor approximation for loss and gradient computations) need to be performed in order to evaluate nonlinear functions. As a result, there is a trade-off between accuracy and privacy. Security measures for the `FedAvg` algorithm needs further studies.

4.4   IMPROVEMENT OF THE FEDAVG ALGORITHM

4.4.1   COMMUNICATION EFFICIENCY

In the implementation of the `FedAvg` algorithm, each participant needs to send a full model weight update to the server during each round of federated training. As modern DNN models can easily have millions of parameters, sending model weights for so many values to a coordinator leads to huge communication costs, which grows with the number of participants and iteration rounds. When there are a large number of participants, uploading model weights from participants to the coordinator becomes the bottleneck of federated learning. To reduce communication costs, some methods are proposed to improve the communication efficiency. One example is Konecný et al. [2016a], in which two strategies for computing model weights are proposed.

- **Sketched updates.** Participants compute a normal model weight update and perform a compression afterward locally. The compressed model weight update is often an unbiased estimator of the true update, meaning they are the same on average. One possible way of performing model weight update compression is using probabilistic quantization. Participants then send the compressed updates to the coordinator.

- **Structured updates.** During the training process, the model weight update is restricted to be of a form that allows for an efficient compression. For example, the model weight may be forced to be sparse or low-rank, or model weight update is computed within a restricted space that can be parameterized using a smaller number of variables. The optimization process then finds the best possible update for this form.

Han et al. [2015] studied DNN model compression proposed a three-stage pipeline for carrying out model weight compression. Firstly, prune the DNN connections by removing redundancy, keeping only the most meaningful connections. Secondly, the weights are quantized so that multiple connections share the same weights, only effective weights are kept. Finally, Huffman coding is applied to take advantage of the biased distribution of effective weights.

When model weights are shared in federated learning, we can use model weight compression to reduce communication costs. Similarly, when gradients are shared in federated learning, we can use gradient compression to bring down communication overhead. One well-known gradient compression method is the deep gradient compression (DGC) approach [Kamp et al., 2018]. DGC employs four methods: namely (1) momentum correction, (2) local gradient clipping, (3) momentum factor masking, and (4) warm-up training. Kamp et al. [2018] applied DGC on image classification, speech recognition, and language modeling tasks. The results showed that DGC can achieve a gradient compression ratio from 270–600 times without compromising model accuracy. Therefore, DGC can be employed to reduce communication bandwidth required for sharing gradients or model weights and facilitate large-scale federated DL or federated learning.

Besides compression, quantization is another efficient method for reducing communication overhead in federated learning [Konecný et al., 2016a, Reisizadeh et al., 2019]. For example, the signSGD-based approach proposed in Chen et al. [2019] has very low per-iteration communication overhead, since it employs one-bit quantization per gradient dimension. Chen et al. [2019] developed a novel gradient correction mechanism that perturbs the local gradients with noise and then applies one-bit quantization, which can also be seen as a special gradient compression scheme.

It is also possible for clients to avoid uploading irrelevant model updates to the server, so as to reduce communication overhead, provided that the training convergence can still be guaranteed [Wang et al., 2019, Hsieh et al., 2017]. For example, Wang et al. [2019] proposed to provide clients with the feedback information regarding the global tendency of model updating. Each client checks whether its local model update aligns with the global tendency and whether it is relevant enough to global model improvement. In this way, each client can decide whether or not it shall upload its local model update to the server. This can be seen as a special case of client selection.

4.4.2   CLIENT SELECTION

In the original work of McMahan et al. [2016b], client selection was recommended to reduce communication cost and time taken for each global training round. However, no approach was proposed for client selection. Nishio and Yonetani [2018] introduced a method for client selection, which consists of two steps. The first step is the resource check step. That is, the coordinator sends queries to a random number of participants to ask about their local resources and the size of data that are relevant to the training task. In the second step, the coordinator uses this information to estimate the time required for each participant to compute model weight update locally, and the time to upload the update. The coordinator then determines which participants to select based on these estimates. The coordinator wants to select as many participants as possible given a specific time budget for one global federated training round.

4.5   RELATED WORKS

The most recent Google workshop on federated learning brought together world-class researchers and practitioners and presented the newest development of federated learning [Google, 2019], such as agnostic federated learning [Mohri et al., 2019], federated transfer learning (see Chapter 6), the incentive mechanism design for federated learning (see Chapter 7), the privacy, security, and fairness aspects of federated learning (see, e.g., Agarwal et al. [2018], Pillutla et al. [2019], Melis et al. [2018], Ma et al. [2019]), as well as a lecture on using the open-source platform TensorFlow Federated [TFF, 2019] for research and deployment of federated learning. We review here some examples of the related studies.

Communication is one of the major challenges of federated learning. In Wang and Joshi [2019], an adaptive communication strategy, termed as AdaComm, was proposed to tackle the problem of random communication delays encountered in federated learning. AdaComm first starts with infrequent model averaging to save communication bandwidth and to deal with random delays, as well as to improve convergence speed. Then it increases the communication frequency in order to achieve better model performance and a low error floor. Wang and Joshi [2019] presented theoretical analysis of the error-runtime trade-off for periodic-averaging SGD algorithm, where each participant performs local updates and their models are averaged periodically (e.g., after every τ iterations). Wang and Joshi [2019] is the first work to analyze the convergence of periodic-averaging SGD in terms of error with respect to wall-clock time, instead of the number of iterations, while taking into account the effect of computation and communication delays on the runtime per iteration. AdaComm is a communication-efficient SGD algorithm for federated learning, which is particularly suitable for mobile applications.

As was discussed in Section 4.3, while the FedAvg algorithm works well for certain non-convex objective functions (also known as cost or loss functions) under IID settings, it could produce unpredictable results for generally non-convex objective functions with non-IID datasets [Goodfellow et al., 2015]. In Xie et al. [2019], a new asynchronous solution was proposed to improve the flexibility and scalability of federated optimization with non-IID training data. The key idea is that the server and the clients conduct model updates asynchronously. The server immediately updates the global model whenever it receives a local model from a client. The communication between the server and the clients is non-blocking. Xie et al. [2019] further analyzed the convergence of their proposed asynchronous approach for a restricted family of non-convex problems under non-IID settings. It was also demonstrated with numerical examples that the asynchronous algorithm enjoys fast convergence and tolerates staleness. Several mixing hyperparameters are introduced to control the trade-off between the convergence rate and variance reduction according to the staleness. However, the hyperparameters of the proposed asynchronous algorithm could be difficult to tune in practice.

The coordinator in HFL is a potential privacy leak and some scholars actually prefer to remove the coordinator. For example, Zantedeschi et al. [2019] considered federated learning under the peer-to-peer architecture, i.e., without the central coordinator, and proposed an optimization procedure that leverages a collaboration graph describing the relationships between the tasks of the participants. The collaboration graph and the ML models are jointly learned. The fully decentralized solution of Zantedeschi et al. [2019] alternates between (i) training nonlinear ML models given the graph in a greedy boosting manner, and (ii) updating the collaboration graph (with controlled sparsity) when given the ML models. Further, the participants exchange messages only with a small number of peers (their direct neighbors in the graph and a few more random participants), thus ensuring the scalability of this fully decentralized solution.

HFL was originally proposed and has been promoted by Google for B2C (business-to-consumer) applications, i.e., mainly for collaborative ML model training employing mobile devices [Konecný et al., 2016b, Yang et al., 2018, Hard et al., 2018], and especially for scenarios with a large number of mobile devices [Bonawitz and Eichner et al., 2019]. While Google is advocating and developing HFL for mobile applications, HFL has been applied to a variety of practical scenarios, particularly to the B2B (business-to-business) applications. For instance, during the Google federated learning workshop [Google, 2019], Prof. Dawn Song from UC Berkeley, talked about applying federated learning for anomaly detection, such as fraud detection [Song, 2019, Hynes et al., 2018], and there are several recent works along this direction, see, e.g., Nguyen et al. [2019] and Preuveneers et al. [2018]. Chapters 8 and 10 will provide more examples of practical applications of HFL.

4.6   CHALLENGES AND OUTLOOK

There are some examples of commercial deployment of HFL systems, such as the one carried out by Google on mobile devices, i.e., the Gboard system [Bonawitz and Eichner et al., 2019]. However, HFL is still in its infancy, and widespread adoption of HFL still faces numerous challenges [Yang et al., 2019, Li, Wen, and He, 2019, Li et al., 2019]. Here, we briefly discuss some of the major challenges.

The first major challenge is the inability to inspect training data, which leads to one of the central problems, namely choosing the hyperparameters of an ML or DL model and setting the optimizers, particularly for training DNN models. It is common to assume that the coordinator or the server has the initial model and knows how to train the model. However, in practice, since we do not collect any data in advance, it is almost impossible to choose the right hyperparameters for a DNN model and set up an optimizer beforehand. Here, the hyperparameters may include the number of layers for a DNN, the number of nodes in each layer of a DNN, the structure of the convolutional neural networks (CNNs), the structure of the recurrent neural networks (RNNs), the output layer of a DNN, and the activation functions, etc. Optimizer options may include which optimizer to use, batch sizes, and learning rates. For example, even the learning rate is difficult to determine since we have no information about the gradient magnitude at each participant. Trying out many different learning rates in production would take time and could worsen development experience. To address this challenge, the simulation based approach suggested by Hartmann [2018, 2019] appears to be promising.

The second challenge is how to effectively motivate companies and organizations to participate in HFL. Traditionally, large companies and organizations have been trying to collect data and create data silos so to be more competitive in the AI age. By joining HFL, other competitors may benefit from such large companies' data, leading to these large companies losing market dominance. As a result, motivating the large companies to adopt HFL could be difficult. To resolve this challenge, we need to devise effective data protection policies, appropriate incentive mechanisms, and business models for HFL.

When applied to mobile devices, it will also be difficult to convince the mobile device owners to allow for their devices to participate in federated learning. Sufficient incentives and benefits shall be demonstrated to the mobile users to draw their interests in offering their mobile devices to join in federated learning, such as potential for better user experience after joining in federated learning.

The third challenge is how to prevent cheating behaviors from participants. It is commonly assumed that the participants are honest. However, in real-life scenarios, honesty only comes under regulations and laws. For example, one participating party may fraudulently claim the number of data points it can contribute for model training and report false testing results of the trained model, in order to gain more rewards. Since we are not able to inspect the datasets of any participants, it is difficult to detect such cheating behaviors. To address this challenge, we need to design a holistic approach for protecting the rights and interests of the honest participants.

To realize the full potential HFL, much research is still needed. In addition to addressing the aforementioned challenges, we need to study mechanisms for managing the training process. For example, since model training and evaluation are carried out locally at each participant, we need to develop new ways to avoid over-fitting and to trigger early-stopping. Another interesting direction is how to handle participants with different reliability levels. For example, some participants may leave during the training process of HFL due to interrupted network connections or other issues. As a result, we need smart solutions to replace the dropped participants with new participants without affecting the training process and model accuracy, especially without affecting the convergence speed of model training. Finally, we need to develop efficient mechanisms to defend against model poisoning attacks (such as targeted backdoor attacks) in federated learning systems.

CHAPTER 5

Vertical Federated Learning

We learned from Chapter 4 that horizontal federated learning (HFL) is applicable to scenarios where participants' datasets share the same feature space but differ in sample spaces. Thus, HFL is convenient to be applied to build applications powered by a massive amount of mobile devices [McMahan et al., 2016a, McMahan and Ramage, 2017]. In those cases, the targets being federated are the individual consumers of applications, which can be considered as B2C (business-to-consumer) paradigm. However, in many practical scenarios, the participants of federated learning are organizations that collected different data features for the same group of people for pursuing different business goals. Those organizations often have strong motivations to cooperate to improve business efficiency, which can be considered as B2B (business-to-business) paradigm.

For example, assume that there is a user has some records in a bank that reflect the user's revenue, expenditure behavior and credit rating. Also, the same user has some other information stored in a e-commerce marketplace retaining the user's online browsing and purchasing history. Although the feature spaces of the two organizations are quite different, they have a close association with each other. For instance, the user's purchasing history may somewhat determine the user's credit rating. Such scenarios are common in real life. Retailers may partner with banks to offer personalized services or products based on the purchasing history and the expenditure behavior of the same user. Hospitals can collaborate with pharmaceutical companies to make use of the medical records of common patients so as to treat chronic diseases and to reduce risks of future hospitalization.

We categorize federated learning on participants whose datasets share the same sample space but differ in feature space as Vertical Federated Learning (VFL). The word "vertical" comes from the term "vertical partition," which is widely used in the context of the traditional tabular view of a database (e.g., columns of a table are vertically partitioned into different groups and each column represents a feature of all samples). In this chapter, we introduce VFL, covering its concept, architecture, algorithms, and open research challenges.

5.1 THE DEFINITION OF VFL

The datasets maintained by different organizations having different business goals usually have different feature spaces, while those organizations may share a large pool of common users. This is illustrated in Figure 5.1. With VFL, also called feature-partitioned federated learning, we can leverage the heterogeneous feature spaces of distributed datasets maintained by those

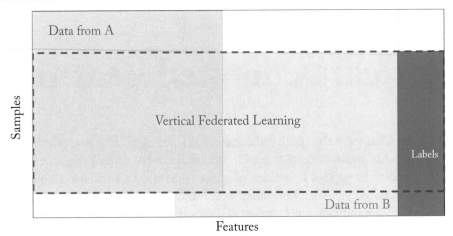

Figure 5.1: Illustration of VFL, a.k.a., feature-partitioned federated learning [Yang et al., 2019].

organizations to build better machine learning (ML) models without exchanging and exposing the private data.

Under the federated learning framework, the identity and the status of each participating party is the same, and the federation helps everyone establish a "commonwealth" strategy, which is why this is called "federated learning." For such a VFL system, we have:

$$\mathcal{X}_i \neq \mathcal{X}_j, \;\; \mathcal{Y}_i \neq \mathcal{Y}_j, \;\; I_i = I_j \;\; \forall \mathcal{D}_i, \mathcal{D}_j, i \neq j, \tag{5.1}$$

where $\mathcal{X}$ and $\mathcal{Y}$ denote the feature space and the label space, respectively. I is the sample ID space, and matrix $\mathcal{D}$ represents the data held by different parties [Yang et al., 2019]. The objective for all parties is to collaboratively build a shared ML model by exploiting all features collected by participating parties.

In VFL settings, there are several underlying assumptions for achieving security and preserving privacy. First, it is assumed that the participants are honest-but-curious. This means that the participants attempt to deduce as much as possible from the information received from other participants, although they abide by the protocol without disturbing it in any way. Since they also intend to build a more accurate model, they do not collude with one another. Second, it is assumed that the information transmission process is safe and reliable enough to defend against attacks. It is further assumed that the communication is lossless without tampering with the intermediate results. A semi-honest third party (STP) may also join the participants to assist the two parities. The STP is independent from both of the two parties. The STP collects the intermediate results to compute the gradients and loss, and distributes the results to each party. The information that the STP receives from the participants is either encrypted or obfuscated. The participants' raw data are not exposed to each other, and each participant only receives the model parameters related to its own features.

Security definition of a VFL system. A VFL system typically assumes honest-but-curious participants. In a two-party case, for example, the two parties are non-colluding and at most one of them are compromised by an adversary. The security definition is that the adversary can only learn data from the party that it corrupted but not data from the other party beyond what is revealed by the input and output. To facilitate the secure computations between the two parties, sometimes a STP is introduced, in which case it is assumed that STP does not collude with either party. MPC provides formal privacy proof for these protocols [Goldreich et al., 1987]. At the end of learning, each party only holds the model parameters associated with its own features, therefore at inference time, the two parties also need to collaborate to generate output.

5.2   ARCHITECTURE OF VFL

For ease of elaboration, we use an example to describe the architecture of VFL. Suppose that Companies A and B would like to jointly train an ML model. Each of them has their own data. In addition, B also has labeled data that the model needs to perform prediction tasks. For user privacy and data security reasons, A and B cannot directly exchange data. In order to ensure data confidentiality during the training process, a third-party collaborator C can be involved. Here, we assume that C is honest and does not collude with A or B, but A and B are honest-but-curious. The trusted third-party C is a legitimate assumption since the role of C can be played by authorities such as governments or replaced by secure computing nodes such as Intel Software Guard Extensions (SGX) [Bahmani et al., 2017]. An example of vertical federated learning architecture is illustrated in Figure 5.2a [Yang et al., 2019, Liu et al., 2019]. The training process of a VFL system typically consists of two parts. It first establishes alignment between entities sharing the same IDs of two parities. Then, the encrypted (or privacy-preserving) training process is conducted on those aligned entities.

Part 1: Encrypted entity alignment. Since the user groups of the two companies A and B are not the same, the system uses an encryption-based user ID alignment technique such as Liang and Chawathe [2004] and Scannapieco et al. [2007] to confirm the common users shared by both parties without A and B exposing their respective raw data. During entity alignment, the system does not expose users who belongs to only one of the two companies, as shown in Figure 5.3.

Part 2: Encrypted model training. After determining the common entities, we can use these common entities' data to train a joint ML model. The training process can be divided into the following four steps (as illustrated in Figure 5.2b).

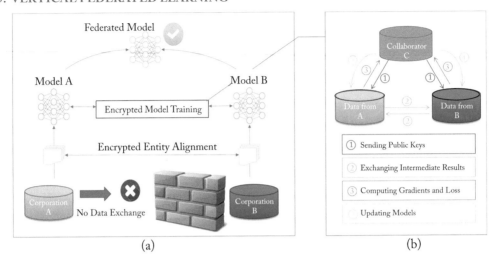

Figure 5.2: Architecture for a vertical federated learning system [Yang et al., 2019].

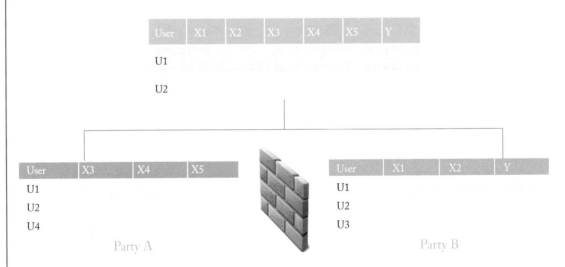

Figure 5.3: Illustration of encrypted entity alignment [Cheng et al., 2019].

- **Step 1:** C creates encryption pairs, and sends the public key to A and B.

- **Step 2:** A and B encrypt and exchange the intermediate results for gradient and loss calculations.

- **Step 3:** A and B compute encrypted gradients and add an *additional mask*, respectively. B also computes the encrypted loss. A and B send encrypted results to C.

- **Step 4:** C decrypts gradients and loss and sends the results back to A and B. A and B unmask the gradients, and update the model parameters accordingly.

5.3   ALGORITHMS OF VFL

In this section, we elaborate two VFL algorithms to help the readers better understand how VFL works.

5.3.1   SECURE FEDERATED LINEAR REGRESSION

The first algorithm is federated linear regression, which was first presented in Yang et al. [2019]. This algorithm exploits homomorphic encryption to protect the privacy of local data belonging to each participating party during the training process of the federated linear regression model. For ease of reference, the notations used in this section are summarized in Table 5.1.

Table 5.1: Notations

η	The learning rate
λ	The regularization parameter
y_i	The label space of party B
x_i^A, x_i^B	Feature space of party A and B, respectively
Θ_A, Θ_B	Local model parameters of party A and B, respectively
u_i^A	Defined as $u_i^A = \Theta_A x_i^A$
u_i^B	Defined as $u_i^B = \Theta_B x_i^B$
$[[d_i]]$	Defined as $[[d_i]] = [[u_i^A]] + [[u_i^B - y_i]]$
$\{x_i^A\}_{i \in \mathcal{D}_A}$	The local dataset of party A
$\{x_i^B, y_i\}_{i \in \mathcal{D}_B}$	The local dataset and labels of party B
$[[\cdot]]$	Additive homomorphic encryption (AHE)
R_A and R_B	The random masks of party A and party B, respectively

To train a linear regression model with gradient descent methods, we need a secure way to compute the model loss and gradients. Given a learning rate η, a regularization parameter λ, a dataset $\{x_i^A\}_{i \in \mathcal{D}_A}$, $\{x_i^B, y_i\}_{i \in \mathcal{D}_B}$, and model parameters Θ_A, Θ_B corresponding to the feature space of x_i^A, x_i^B, respectively, the optimization problem for model training can be written as:

$$\min_{\Theta_A, \Theta_B} \sum_i \left\| \Theta_A x_i^A + \Theta_B x_i^B - y_i \right\|^2 + \frac{\lambda}{2} \left(\|\Theta_A\|^2 + \|\Theta_B\|^2 \right). \tag{5.2}$$

Let $u_i^A = \Theta_A x_i^A$, $u_i^B = \Theta_B x_i^B$, the encrypted loss is:

$$[[\mathcal{L}]] = \left[\left[\sum_i \left(\left(u_i^A + u_i^B - y_i\right)\right)^2 + \frac{\lambda}{2}\left(\|\Theta_A\|^2 + \|\Theta_B\|^2\right)\right]\right], \tag{5.3}$$

where the additive homomorphic encryption operation is denoted as $[[\cdot]]$. Let $[[\mathcal{L}_A]] = [[\sum_i (u_i^A)^2 + \frac{\lambda}{2}\|\Theta_A\|^2]]$, $[[\mathcal{L}_B]] = [[\sum_i (u_i^B - y_i)^2 + \frac{\lambda}{2}\|\Theta_B\|^2]]$, and $[[\mathcal{L}_{AB}]] = 2\sum_i [[u_i^A (u_i^B - y_i)]]$, then

$$[[\mathcal{L}]] = [[\mathcal{L}_A]] + [[\mathcal{L}_B]] + [[\mathcal{L}_{AB}]]. \tag{5.4}$$

Similarly, let $[[d_i]] = [[u_i^A]] + [[u_i^B - y_i]]$. Then, the gradients of the loss function with respect to the training parameters are given by

$$\left[\left[\frac{\partial \mathcal{L}}{\partial \Theta_A}\right]\right] = 2\sum_i [[d_i]] x_i^A + [[\lambda \Theta_A]] \tag{5.5}$$

$$\left[\left[\frac{\partial \mathcal{L}}{\partial \Theta_B}\right]\right] = 2\sum_i [[d_i]] x_i^B + [[\lambda \Theta_B]]. \tag{5.6}$$

Note that party A and party B are able to compute u_i^A and u_i^B using only their respective local information. However, the term d_i involves both u_i^A and $u_i^B - y_i$. It cannot be computed by either party alone. As a result, party A and party B should collaboratively compute d_i while preserving the privacy of u_i^A and $u_i^B - y_i$ against the other party. In the homomorphic encryption setting, to prevent party A and party B from peeking into $u_i^B - y_i$ and u_i^A, respectively, $u_i^B - y_i$ and u_i^A are encrypted via the public key held by a third party C. The third party C in the loop is mainly response for decrypting encrypted information passed from party A and party B, and orchestrating the training and inference processes.

Introducing a third party into the loop is not always feasible to many practical scenarios where the legitimacy and accountability of the third party are hard to be guaranteed. Secure multi-party computation techniques such as secret sharing can be applied to remove the third party and decentralize federated learning. We refer interested readers to Mohassel and Zhang [2017] for details. Here, we continue the elaboration with the architecture having a third party.

The Training Process

We summarize the detailed steps of training the federated linear regression model in Table 5.2. During entity alignment and model training, data owned by parties A and B are stored locally, and the interactions in model training do not lead to data privacy leakage. Note that potential information leakage to party C may or may not be considered to be privacy violation since party C is a trusted party. To further prevent party C from learning information about parties A or B in this case, parties A and B can further hide their gradients from party C by adding encrypted random masks.

Table 5.2: Training steps for secure linear regression

	Party A	Party B	Party C
Step 1	Initializes Θ_A	Initializes Θ_B	Creates an encryption key pair and sends public key to A and B
Step 2	Computes $[[u_i^A]]$, $[[\mathcal{L}_A]]$ and sends to B	Computes $[[u_i^B]]$, $[[d_i^B]]$, $[[\mathcal{L}]]$, and sends $[[d_i^B]]$ to A, and sends $[[\mathcal{L}]]$ to C	
Step 3	Initializes R_A, computes $[[\frac{\partial \mathcal{L}}{\partial \Theta_A}]] + [[R_A]]$ and sends to C	Initializes R_B, computes $[[\frac{\partial \mathcal{L}}{\partial \Theta_B}]] + [[R_B]]$ and sends to C	Decrypts $[[\mathcal{L}]]$ and sends $[[\frac{\partial \mathcal{L}}{\partial \Theta_A}]] + R_A$ to A, $[[\frac{\partial \mathcal{L}}{\partial \Theta_B}]] + R_B$ to B
Step 4	Updates Θ_A	Updates Θ_B	
What is obtained?	Θ_A	Θ_B	

The training protocol shown in Table 5.2 does not reveal any information to C, because all C learns are the masked gradients, and the randomness and secrecy of the masked matrix are guaranteed [Du et al., 2004]. In the above protocol, party A learns its gradient at each step, but this is not enough for A to learn any information about B according to Equation (5.5), because the security of scalar product protocol is well-established based on the inability of definitively solving for more than n unknowns with only n equations [Du et al., 2004, Vaidya and Clifton, 2002]. Here, we assume the number of samples N_A is much greater than n_A, where n_A is the number of features. Similarly, party B cannot learn any information about A. Therefore, the security of the protocol is proven.

Note that we have assumed both parties to be semi-honest. If a party is malicious and cheats the system by faking its input, for example, party A submits only one non-zero sample with only one non-zero feature, it can tell the value of u_i^B for that feature of that sample. It still cannot compromise x_i^B or Θ_B though, and the deviation will distort results for the next iteration, thereby alerting the other party who can then terminate the learning process in response. At the end of the training process, each party remains oblivious about the data structure of the other party, and it obtains the model parameters associated only with its own features.

Because the loss and gradients each party receives are exactly the same as the loss and gradients they would receive if jointly building a model with data gathered at one place without privacy constraints, this collaboratively trained model is lossless and its optimality is guaranteed.

The efficiency of the model depends on the communication overhead and the computation overhead incurred for encrypting the data. In each iteration, the information sent between A and

B increases with the number of overlapping samples. Therefore, the efficiency of this algorithm can be further improved by adopting distributed parallel computing techniques.

The Inference Process

At inference time, the two parties need to collaboratively compute the inference results, as the steps summarized in Table 5.3. During inference, the data belonging to each party would not be exposed to the other.

Table 5.3: Inference steps for secure linear regression

	Party A	Party B	Evaluator C
Step 0			Sends user ID i to A and B
Step 1	Computes u_i^A and sends to C	Computes u_i^B and sends to C	Computes the result of $u_i^A + u_i^B$

5.3.2   SECURE FEDERATED TREE-BOOSTING

The second example is secure federated tree-boosting (termed as SecureBoost for short), which was first studied in Cheng et al. [2019], in the setting of VFL. This work proves that SecureBoost is as accurate as other non-federated gradient tree-boosting algorithms that require data being collected at one central place. That is, SecureBoost provides the same level of accuracy as its non-privacy-preserving variants, while at the same time, reveals no information about each private data owner. Note that, in the description of this example, the coordinator corresponds to the so-called *active party* originally defined in Cheng et al. [2019], which has training data with labels. There are also passive parties that have training data without labels.

Secure Entity Alignment

Similar to federated secure linear regression described in Section 5.3.1, SecureBoost consists of two major steps. First, it aligns the data under the privacy constraint. Second, it collaboratively learns a shared gradient-tree boosting model, while keeping all the training data secret over multiple private parties.

The first step in the SecureBoost framework is entity alignment, which is to find a common set of data samples (i.e., the common users) among all participating parties so as to build a joint ML model. When the data are vertically partitioned over multiple parties, different parties hold different but partially overlapping users' data. The common users may be identified by their unique user IDs. In particular, we can align the data samples under an encryption scheme by using the privacy-preserving protocol for inter-database intersections, see, e.g., Liang and Chawathe [2004].

Review of XGBoost

After aligning the data across different parties under the privacy constraints, we now consider the problem of jointly building the tree ensemble model over multiple parties without violating privacy through VFL. To achieve this goal, there are three key questions that need to be answered.

- How can each party compute an updated model based on its local data without reference to class labels?

- How can the coordinator aggregate all the updated models and obtain a new global model?

- How to share the updated global model among all parties without leaking any private information at the inference time?

To help find answers to these questions, we first take a quick review of the tree ensemble model, XGBoost [Chen and Guestrin, 2016], in the non-federated setting.

Given a data set $\mathbf{X} \in \mathbb{R}^{n \times d}$ with n samples and d features, XGBoost predicts the output by using K regression trees:

$$\hat{y}_i = \sum_{k=1}^{K} f_k(\mathbf{x}_i), \ \forall \mathbf{x}_i \in \mathbb{R}^d, \ i = 1, 2, \ldots, n. \tag{5.7}$$

To avoid bogging down into the mathematical details of tree boosting, we bypass the derivation of the loss function for learning the set of regression trees used in Equation (5.7). Instead, we introduce the training rules with pain words, hoping that it is more accessible. The objective of learning the regression tree ensemble is to find a best set of trees that provides small classification loss, as well as low model complexity. In gradient tree boosting, this objective is approached by optimizing the loss (e.g., squared loss or Taylor approximation of a loss function) between label and prediction iteratively. In each iteration, we try to add a new tree that reduces the loss as much as possible while do not introduces much complexity. Hence, the objective function at tth iteration can be written as

$$\mathcal{L}^{(t)} \triangleq \sum_{i=1}^{n} \left[l\left(y_i, \hat{y}_i^{(t-1)}\right) + g_i f_t(\mathbf{x}_i) + \frac{1}{2} h_i f_t^2(\mathbf{x}_i) \right] + \Omega(f_t), \tag{5.8}$$

where g_i and h_i are two groups of independent variables, and $\Omega(f_t)$ are the complexity of the new tree. This indicates that we only need to find a new tree that can optimize this objective function, and it should be easy to find a optimal tree given this objective function with the optimal score noted as obj_t.

Hence, the construction of a newly added regression is a process to build a tree that minimize obj_t, i.e., from the depth of 0, deciding the split of each leaf node until reaching the maximum depth. Now the question lays on how to decide a optimal split of a leaf node at each level of the tree. The performance of a "split" is measured by the split gain, which can be calculated from the aforementioned variables g_i and h_i. We have the following observations.

(i) The evaluation of split candidates and the calculation of the optimal weight of a leaf node only depend on the variables g_i and h_i.

(ii) The class label is needed for the calculation of g_i and h_i, and is easy to recover the class label from g_i and h_i once we obtain the value of $y_i^{(t-1)}$ from the $(t-1)$th iteration.

Algorithm 5.1 Aggregate Dencrypted Gradient Statistics (adapted from Cheng et al. [2019])

Input: I, the instance space of the current node;
Input: d, feature dimension;
Input: $\{[[g_i]], [[h_i]]\}_{i \in I}$.
Output: $\mathbf{G} \in \mathbb{R}^{d \times l}, \mathbf{H} \in \mathbb{R}^{d \times l}$
1: **for** $k = 0 \to d$ **do**
2:    Propose $S_k = \{s_{k1}, s_{k2}, ..., s_{kl}\}$ by percentiles on feature k
3: **end for**
4: **for** $k = 0 \to d$ **do**
5:    $\mathbf{G}_{kv} = \sum_{i \in \{i | s_{k,v} \geq x_{i,k} > s_{k,v-1}\}} [[g_i]]$
6:    $\mathbf{H}_{kv} = \sum_{i \in \{i | s_{k,v} \geq x_{i,k} > s_{k,v-1}\}} [[h_i]]$
7: **end for**

The Training Process of SecureBoost

We know from the above observations that each party can determine the local optimal split independently with only its local data once it obtains g_i and h_i. The optimal split can be found if the split gain can be calculated for every possible splits, by using the sum of groups of g_i and h_i.

In order to keep g_i and h_i confidential to avoid privacy leakage, g_i and h_i shall be encrypted before being sent to the other parties. As an example, we show here how to calculate the candidate split gain with encrypted g_i and h_i using the additive homomorphic encryption scheme [Paillier, 1999].

With the additive homomorphic encryption scheme, the split gain for every split candidate can be computed by the sum of groups of ciphertexts of g_i and h_i, respectively. Therefore, the best split at each party can be found by evaluating all the possible split gains in the coordinator that can hence apply a global optimal split.

However, this solution is not efficient since it requires the transmission for all possible split candidates, which incurs enormous communication overhead. To construct a boosting-tree with lower communication cost, we can take advantage of the approximate framework proposed in Chen and Guestrin [2016], where the detailed calculation is shown in Algorithm 5.1.

Algorithm 5.2 Find Optimal Split (adapted from Cheng et al. [2019])

Input: I, instance space of current node
Input: $\{G^i, H^i\}_{i=1}^{m}$, aggregated encrypted gradient statistics from m parties
Output: Partition current instance space according to the selected attribute's value

1: **The Coordinator executes:**
2: $g \leftarrow \sum_{i \in I} g_i, \quad h \leftarrow \sum_{i \in I} h_i$

3: **Enumerate over all parties:**
4: **for** $i = 0 \rightarrow m$ **do**
5:   Enumerate over all features
6:   **for** $k = 0 \rightarrow d_i$ **do**
7:     $g_l \leftarrow 0, h_l \leftarrow 0$
8:     Enumerate over all threshold value
9:     **for** $v = 0 \rightarrow l_k$ **do**
10:       Get decrypted values $D(G_{kv}^i)$ and $D(H_{kv}^i)$
11:       $g_l \leftarrow g_l + D(G_{kv}^i), h_l \leftarrow h_l + D(H_{kv}^i)$
12:       $g_r \leftarrow g - g_l, h_r \leftarrow h - h_l$
13:       $score \leftarrow max(score, \frac{g_l^2}{h_l+\lambda} + \frac{g_r^2}{h_r+\lambda} - \frac{g^2}{h+\lambda})$
14:     **end for**
15:   **end for**
16: **end for**
17: Return k_{opt} and v_{opt} to the corresponding party i when we obtain the max score.

18: **Party i executes (for many parties in parallel):**
19: Determine the selected attribute's value according to k_{opt} and v_{opt} and partition current instance space.
20: Record the selected attribute's value and return [record id, I_L] back to the collaborator.

21: **The Coordinator executes:**
22: Split current node according to I_L and associate current node with [party id, record id].

With Algorithm 5.1, for each party, instead of computing $[[g_l]]$ and $[[h_l]]$ directly, it maps the features into buckets and then aggregates the encrypted gradient statistics based on the buckets. In this way, the coordinator only needs to collect the aggregated encrypted gradient statistics from all parties. As a result, it can determine the globally optimal split more efficiently, as described in Algorithm 5.2.

After the coordinator obtains the global optimal split, represented as [party id (i), feature id (k), threshold id (v)], it returns the feature id k and threshold id v to the corresponding party

i. Party i decides the value of the selected attribute based on the values of k and v. Then, it partitions the current instance space according to the value of the selected attribute. In addition, it builds a lookup table locally to record the value of the selected attribute, [feature, threshold value]. After that, it returns the index of the record and the instance space of left side nodes after the split (I_L) back to the active party. The active party splits the current node according to the received instance space and associates the current node with [party id, record id], until a stopping criterion or the maximum depth is reached. All the leaf nodes are stored inside the active party.

In summary, the step-by-step training process of SecureBoost can be concluded as follows.

- **Step 1:** Starting from the active party, it first calculates g_i and h_i, $i \in \{1, \ldots, N\}$, and encrypts it using AHE. The active party sends g_i and h_i, $i \in \{1, \ldots, N\}$, to all the passive parties.

- **Step 2:** For each passive party, it maps the features into buckets and then aggregates the encrypted gradient statistics based on the those buckets. The results are sent to the active party.

- **Step 3:** The active party decrypts the aggregated result, and determines the global optimal split according to Algorithm 5.2, and returns k_{opt} and v_{opt} to the corresponding passive party.

- **Step 4:** The passive party determines the attribute's value according to k_{opt} and v_{opt} received from the active party, and returns the corresponding records to the active party.

- **Step 5:** The active party splits the current node according to the received instance space (I_L) and associates the current node with [party-id, record-id].

- **Step 6:** Repeat Steps 2–5 iteratively until the stopping criterion is reached.

The Inference Process of SecureBoost

For federated model inference, we are able to classify a new sample even though its features are privately distributed on different parties. Since all leaf nodes are stored in the active party, the inference process should be coordinated by the active party with the information from other passive parties, which have party-specific lookup table consisting of [feature, threshold value]. The inference process is simply recursive steps as follows.

- **Step 1:** The active party refers to the owner (i.e., party-id) of the current node with the related feature-threshold tuple (i.e., record-id).

- **Step 2:** The party found by Step 1 compares the value of the corresponding attribute with the threshold from the lookup table, decides which child node to retrieve, and returns the decision to the active party.

- **Step 3:** The active party goes to the child node according to the decision received.

- **Step 4:** Repeat Step 1 – Step 3 until a leaf node is reached.

In each step, the active party only reacts to the retrieval of tree node and its corresponding party-id and record-id. The actual attribute values are only exposed to the parties that owns the corresponding attributes. Therefore, the federated inference is not only private but also lossless.

5.4 CHALLENGES AND OUTLOOK

VFL enables participants to build a shared model based on data with heterogeneous features in a privacy-preserving manner. Unlike HFL in which a common model is shared by all participants, in VFL the model is partitioned into multiple components each maintained by a participant with relevant but different data features. Thus, participants in VFL have a closer interdependent relationship with each other. More specifically, the training of each model component must follow a certain computation order specified by the underlying VFL algorithm. In other words, participants have dependent computations and need to frequently interact with each other to exchange the intermediate results.

Therefore, VFL is vulnerable to communication failures and thus requires reliable and efficient communication mechanisms. Transferring the intermediate results from one participant to another can be expensive, as long-haul connections must be established between two participants located in different geographical regions. Such slow data transfer, in turn, results in inefficient utilization of computing resources, as a participant cannot start training until it has received all the required intermediate results. To address this issue, we may need to design a streaming communication mechanism that can judiciously schedule the training and communication of each participant to offset the data transfer delay. A fault-tolerant mechanism should also be designed to prevent the VFL process from crashing in the middle of training.

Currently, on the setting of federated learning, most works proposed to reduce information leakage or prevent malicious attacks are applied in HFL. As VFL typically requires a closer and more direct interaction between participants, flexible secure protocols that can meet secure requirements of each participant are needed. Previous works [Baldimtsi et al., 2018, Bost et al., 2015] have demonstrated that different secure tools are optimal for different types of computations, e.g., garbled circuits give efficient comparisons whereas secret sharing and homomorphic encryption yield efficient arithmetic function evaluation. We may explore hybrid strategies for conversion among secure techniques, aiming to achieve locally optimal performance for each part of the computation. In addition, efficient secure entity alignment is also worth being explored since it is a crucial preprocessing component of VFL.

CHAPTER 6

Federated Transfer Learning

We have discussed horizontal federated learning (HFL) and vertical federated learning (VFL) in Chapters 4 and 5, respectively. HFL requires all participating parties share the same feature space while VFL require parties share the same sample space. In practice, however, we often face situations in which there are not enough shared features or samples among the participating parties. In those cases, one can still build a federated learning model combined with transfer learning that transfers knowledge among the parties to achieve better performance. We refer to the combination of federated learning and transfer learning as Federated Transfer Learning (FTL). In this chapter, we provide a formal definition of FTL and discuss the differences between FTL and traditional transfer learning. We then introduce a secure FTL framework proposed in Liu et al. [2019], and conclude this chapter with a summary of the challenges and open issues.

6.1 HETEROGENEOUS FEDERATED LEARNING

Both HFL and VFL require all participants share either the same feature space or the same sample space in order to build an effective shared machine learning (ML) model. In more practical scenarios, however, datasets maintained by participants may be highly heterogeneous in one way or the other.

- Datasets may share only a handful of samples and features.

- Distributions among those datasets could be quite different.

- The size of those datasets could vary greatly.

- Some participants may only have data with no or limited labels.

To address these issues, federated learning can be combined with transfer learning techniques [Pan and Yang, 2010] to enable a broader range of businesses and applications that have only small data (few overlapping samples and features) and weak supervision (few labels) to build effective and accurate ML models while complying with data privacy and security law [Yang et al., 2019, Liu et al., 2019]. We refer to the combination of federated learning and transfer learning as FTL, which deals with problems that exceed the scope of the existing HFL and VFL settings.

6.2   FEDERATED TRANSFER LEARNING

Transfer learning is a learning technique to provide solutions for cross-domain knowledge transfer. In many applications, we only have a small amount of labeled data or weak supervision such that ML models cannot be built reliably [Pan and Yang, 2010]. In such situations, we can still build high-performance ML models by leveraging and adapting models from similar tasks or domains. In recent years, there have been an increasing number of research works on applying transfer learning to various fields ranging from image classification [Zhu et al., 2011] to natural language understanding and sentiment analysis [Li et al., 2017, Pan et al., 2010].

The essence of transfer learning is to find the invariant between a resource-rich source domain and a resource-scarce target domain, and exploit that invariant to transfer knowledge from source domain to target domain. Based on approaches used to conduct transfer learning, Pan and Yang [2010] divides transfer learning into mainly three categories: (i) instance-based transfer, (ii) feature-based transfer, and (iii) model-based transfer. FTL extends the traditional transfer learning to the privacy-preserving distributed machine learning (DML) paradigm. Here, we describe how these three categories of transfer learning techniques can be applied to HFL and VFL, respectively.

- **Instance-based FTL.** For HFL, data of participating parties are typically drawn from different distributions, which may lead to the poor performance of ML models trained on those data. Participating parties can selectively pick or re-weight training data samples to relieve the distribution difference such that the objective loss function can be optimally minimized. For VFL, participating parties may have quite different business objectives. Thus, aligned samples and some of their features may have a negative impact on the federated transfer learning, which is referred to as negative transfer [Pan and Yang, 2010]. In this scenario, participating parties can selectively choose features and samples for avoiding negative transfer.

- **Feature-based FTL.** Participating parties collaboratively learn a common feature representation space, in which the distribution and semantic difference among feature representations transformed from raw data can be relieved and such that knowledge can be transferable across different domains. For HFL, the common feature representation space can be learned through minimizing the maximum mean discrepancy (MMD) [Pan et al., 2009] among samples of participating parties. While for VFL, the common feature representation space can be learned through minimizing the distance between representations of aligned samples belonging to different parties.

- **Model-based FTL.** Participating parties collaboratively learn shared models that can benefit for transfer learning. Alternatively, participating parties utilize pre-trained models as the whole or part of the initial models for a federated learning task. HFL is a kind of model-based FTL since during training a shared global model is being learned based on data of all parties, and that shared global model is served as a pre-trained model to be fine-tuned by each party in each communication round [McMahan et al., 2016a]. For VFL, predictive models can be learned from aligned samples for inferring missing features and labels (i.e., the blank spaces in Figure 1.4). Then, the enlarged training samples can be used to train a more accurate shared model.

Formally, FTL aims to provide solutions for situations when:

$$\mathcal{X}_i \neq \mathcal{X}_j, \ \mathcal{Y}_i \neq \mathcal{Y}_j, \ \mathcal{I}_i \neq \mathcal{I}_j, \ \forall \mathcal{D}_i, \mathcal{D}_j, i \neq j, \tag{6.1}$$

where $\mathcal{X}_i$ and $\mathcal{Y}_i$ denote the feature space and the label space of the ith party, respectively; $\mathcal{I}_i$ stands for the sample space, and matrix $\mathcal{D}_i$ represents the dataset held by the ith party [Yang et al., 2019]. The objective is to predict labels for newly incoming samples or existing unlabeled samples as accurately as possible.

In Section 6.3, we will introduce a secure feature-based FTL framework proposed by Liu et al. [2019] that helps predict labels for target domain by exploiting knowledge transferred from source domain.

From the technical perspective, FTL differs from traditional transfer learning mainly in the following two ways.

- FTL builds models based on data distributed among multiple parties, and the data belonging to each party cannot be gathered together or exposed to other parties. Traditional transfer learning has no such constraint.

- FTL requires the preservation of user privacy and the protection of data (and model) security, which is not a significant concern in traditional transfer learning.

FTL brings traditional transfer learning into the privacy-preserving DML paradigm. Therefore, we should define the security that a FTL system must guarantee.

Definition 6.1   Security definition of a FTL system. An FTL system typically involves two parties, namely the source domain party and the target domain party. A multi-party FTL system can be regarded as a combination of multiple two-party FTL subsystems. It is assumed that both parties are honest-but-curious. That is, all parties in the federation follow the federation protocols and rules but they will try to deduce information from data received. Consider a threat model with a semi-honest adversary who can corrupt at most one of the two parties of a two-party FTL system. For a protocol P performing $(O_A, O_B) = P(I_A, I_B)$, where O_A and O_B are party A's and party B's respective outputs, and I_A and I_B are their respective inputs, P is secure against party A if there exists an infinite number of (I'_B, O'_B) pairs such that $(O_A, O'_B) = P(I_A, I'_B)$. Such a security definition has been adopted in Du et al. [2004]. It provides a practical solution to control information disclosure as compared to complete zero knowledge security.

6.3   THE FTL FRAMEWORK

In this section, we introduce a secure feature-based FTL framework proposed by Liu et al.
[2019]. Figure 6.1 illustrates this FTL framework in which a predictive model learned from
feature representations of aligned samples belonging to party A and party B is utilized to predict
labels for unlabeled samples of party B.

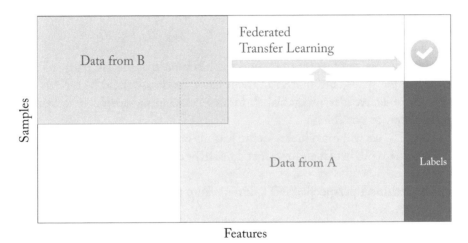

Figure 6.1: Illustration of FTL [Yang et al., 2019]. A predictive model learned from feature
representations of aligned samples belonging to party A and party B is utilized to predict labels
for unlabeled samples of party B.

Consider a source domain party A with dataset $\mathcal{D}_A := \{(x_i^A, y_i^A)\}_{i=1}^{N_A}$ where $x_i^A \in R^a$ and
$y_i^A \in \{+1, -1\}$ is the ith label, a target domain party B with dataset $\mathcal{D}_B := \{x_j^B\}_{j=1}^{N_B}$ where
$x_j^B \in R^a$. $\mathcal{D}_A, \mathcal{D}_B$ are separately held by two private parties and cannot be exposed to each other.
We also assume that there exists a limited set of co-occurring samples $\mathcal{D}_{AB} := \{(x_i^A, x_i^B)\}_{i=1}^{N_{AB}}$ and
a small set of labels for B's data in party A: $\mathcal{D}_c := \{(x_i^B, y_i^A)\}_{i=1}^{N_c}$, where N_c is the number of
available target labels.

Without loss of generality, we assume all labels are in party A, but all the description here
can be adapted to the case where labels exist in party B. One can find the commonly shared
sample ID set in a privacy-preserving setting by masking data IDs with encryption techniques
such as the RSA scheme. Here, we assume that A and B already found or both know their
commonly shared sample IDs. Given the above setting, the objective is for the two parities to
collaboratively build a transfer learning model to predict labels for the target-domain party B as
accurately as possible without exposing data to each other.

In recent years, DNNs have been widely adopted in transfer learning to find the implicit transfer mechanism [Oquab et al., 2014]. Here, we explore a general scenario in which

hidden representations of A and B are produced by two neural networks $u_i^A = Net^A(x_i^A)$ and $u_i^B = Net^B(x_i^B)$, where $u^A \in \mathbb{R}^{N_A \times d}$ and $u^B \in \mathbb{R}^{N_B \times d}$, d is the dimension of the hidden representation layer. Figure 6.2 illustrates the architecture of two neural networks.

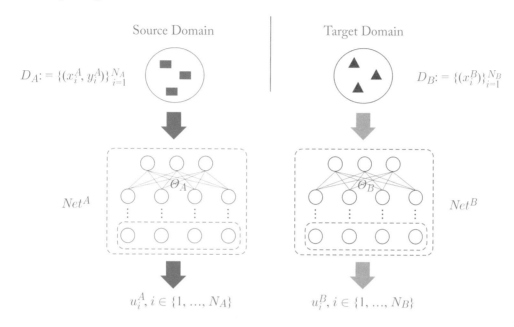

Figure 6.2: The architecture of neural networks of source and target domains.

To label the data in the target domain, a general approach is to introduce a prediction function $\varphi(u_j^B) = \varphi(u_1^A, y_1^A \ldots u_{N_A}^A, y_{N_A}^A, u_j^B)$. For example, Shu et al. [2015] used a translator function, $\varphi(u_j^B) = \frac{1}{N_A} \sum_i^{N_A} y_i^A u_i^A (u_j^B)'$. We can then write the optimization problem for model training using the available labeled dataset as:

$$\min_{\Theta^A, \Theta^B} \mathcal{L}_1 = \sum_i^{N_c} \ell_1 \left(y_i^A, \varphi \left(u_i^B \right) \right), \qquad (6.2)$$

where Θ^A, Θ^B are training parameters of Net^A and Net^B, respectively. Let L_A and L_B be the number of layers for Net^A and Net^B, respectively. Then, $\Theta^A = \{\theta_l^A\}_{l=1}^{L_A}$, $\Theta^B = \{\theta_l^B\}_{l=1}^{L_B}$ where θ_l^A and θ_l^B are the training parameters for the lth layer. ℓ_1 denotes the loss function. For logistic loss, $\ell_1(y, \varphi) = \log(1 + e^{-y\varphi})$.

In addition, we also aim to minimize the alignment loss between A and B. That is,

$$\min_{\Theta^A, \Theta^B} \mathcal{L}_2 = -\sum_i^{N_{AB}} \ell_2 \left(u_i^A, u_i^B \right), \qquad (6.3)$$

where ℓ_2 denotes the alignment loss which can be represented as $-u_i^A(u_i^B)'$ or $||u_i^A - u_i^B||_F^2$. For simplicity, we assume it can be expressed in the form $\ell_2(u_i^A, u_i^B) = \ell_2^A(u_i^A) + \ell_2^B(u_i^B) + \kappa u_i^A(u_i^B)'$, where κ is a constant.

The final objective function for model training is:

$$\mathcal{L} = \mathcal{L}_1 + \gamma \mathcal{L}_2 + \frac{\lambda}{2}\left(\mathcal{L}_3^A + \mathcal{L}_3^B\right), \tag{6.4}$$

where γ and λ are the weight parameters, and $\mathcal{L}_3^A = \sum_l^{L_A} ||\theta_l^A||_F^2$, $\mathcal{L}_3^B = \sum_l^{L_B} ||\theta_l^B||_F^2$ are the regularization terms.

The next step is to obtain the gradients for updating Θ^A, Θ^B through back propagation. For $i \in \{A, B\}$, we have

$$\frac{\partial \mathcal{L}}{\partial \theta_l^i} = \frac{\partial \mathcal{L}_1}{\partial \theta_l^i} + \gamma \frac{\partial \mathcal{L}_2}{\partial \theta_l^i} + \lambda \theta_l^i. \tag{6.5}$$

Under the condition that A and B shall not expose their raw data, privacy-preserving approaches need to be developed to compute the loss in Equation (6.4) and the gradients in Equation (6.5). We describe two secure federated transfer learning approaches at high level for computing Equations (6.4) and (6.5). One is based on homomorphic encryption [Acar et al., 2018] and the other is based on secret sharing. In both approaches, we adopt second-order Taylor approximation for computing (6.4) and (6.5).

6.3.1   ADDITIVELY HOMOMORPHIC ENCRYPTION

Additively homomorphic encryption [Acar et al., 2018] and polynomial approximations have been widely used for privacy-preserving ML. The trade-offs between efficiency and privacy by adopting such approximations have been discussed in detail in Aono et al. [2016], Kim et al. [2018], and Phong et al. [2018]. Applying Equations (6.4) and (6.5), and additively homomorphic encryption (denoted as $[[\cdot]]$, see also Section 2.4.2), we obtain the privacy preserved loss function and the corresponding gradients for the two domains as:

$$[[\mathcal{L}]] = [[\mathcal{L}_1]] + [[\gamma \mathcal{L}_2]] + \left[\left[\frac{\lambda}{2}\left(\mathcal{L}_3^A + \mathcal{L}_3^B\right)\right]\right], \tag{6.6}$$

$$\left[\left[\frac{\partial \mathcal{L}}{\partial \theta_l^B}\right]\right] = \left[\left[\frac{\partial \mathcal{L}_1}{\partial \theta_l^B}\right]\right] + \left[\left[\gamma \frac{\partial \mathcal{L}_2}{\partial \theta_l^B}\right]\right] + [[\lambda \theta_l^B]] \tag{6.7}$$

$$\left[\left[\frac{\partial \mathcal{L}}{\partial \theta_l^A}\right]\right] = \left[\left[\frac{\partial \mathcal{L}_1}{\partial \theta_l^A}\right]\right] + \left[\left[\gamma \frac{\partial \mathcal{L}_2}{\partial \theta_l^A}\right]\right] + [[\lambda \theta_l^A]]. \tag{6.8}$$

Let $[[\cdot]]_A$ and $[[\cdot]]_B$ be homomorphic encryption operators with public keys from A and B, respectively. Let $[[(\frac{\partial \mathcal{L}}{\partial \theta_l^B})^A]]_A$ be a set of intermediate components computed and encrypted by party A for calculating $[[\frac{\partial \mathcal{L}}{\partial \theta_l^B}]]$. Let $[[(\frac{\partial \mathcal{L}}{\partial \theta_l^A})^B]]_B$, $[[\mathcal{L}^B]]_B$ be a set of intermediate components computed and encrypted by party B for calculating $[[\frac{\partial \mathcal{L}}{\partial \theta_l^A}]]$, $[[\mathcal{L}]]$, respectively.

Note that we exclude mathematical details of loss and gradients calculation, and focus on the collaboration between participating parties. We refer interested readers to Liu et al. [2019] for detailed explaination of the secure FTL framework.

6.3.2   THE FTL TRAINING PROCESS

With Equations (6.6), (6.7), and (6.8), we can now design a federated algorithm for training the FTL model. The training process contains the following steps.

- **Step 1**: Party A and party B initialize and run their independent neural networks Net^A and Net^B locally to obtain hidden representations u_i^A and u_i^B.

- **Step 2**: Party A computes and encrypts a list of intermediate components, denoted as $[[(\frac{\partial \mathcal{L}}{\partial \theta_l^B})^A]]_A$ and sends them to B to assist with the calculations of gradients $\frac{\partial \mathcal{L}}{\partial \theta_l^B}$. While party B computes and encrypts a list of intermediate components, denoted as $[[(\frac{\partial \mathcal{L}}{\partial \theta_l^A})^B]]_B$, $[[\mathcal{L}^B]]_B$, and sends them to A to assist with the calculations of gradients $\frac{\partial \mathcal{L}}{\partial \theta_l^A}$ and loss $\mathcal{L}$.

- **Step 3**: Based on $[[(\frac{\partial \mathcal{L}}{\partial \theta_l^A})^B]]_B$ and $[[\mathcal{L}^B]]_B$ received, party A computes $[[\frac{\partial \mathcal{L}}{\partial \theta_l^A}]]_B$ and $[[\mathcal{L}]]_B$ via (6.6) and (6.8). Then party A creates random mask m^A and add it to $[[\frac{\partial \mathcal{L}}{\partial \theta_l^A}]]_B$ to obtain $[[\frac{\partial \mathcal{L}}{\partial \theta_l^A} + m^A]]_B$. Party A sends $[[\frac{\partial \mathcal{L}}{\partial \theta_l^A} + m^A]]_B$ and $[[\mathcal{L}]]_B$ to B. Based on $[[(\frac{\partial \mathcal{L}}{\partial \theta_l^B})^A]]_A$ received, party B computes $[[\frac{\partial \mathcal{L}}{\partial \theta_l^B}]]_A$ via Equation (6.7). Then party B creates random mask m^B and add it to $[[\frac{\partial \mathcal{L}}{\partial \theta_l^B}]]_A$ to obtain $[[\frac{\partial \mathcal{L}}{\partial \theta_l^B} + m^B]]_A$. Party B sends $[[\frac{\partial \mathcal{L}}{\partial \theta_l^B} + m^B]]_A$ to A.

- **Step 4**: Party A decrypts $\frac{\partial \mathcal{L}}{\partial \theta_l^B} + m^B$ and sends it to B. While party B decrypts $\frac{\partial \mathcal{L}}{\partial \theta_l^A} + m^A$ and $\mathcal{L}$, and sends them to A.

- **Step 5**: Party A and party B remove random masks and obtain gradients $\frac{\partial \mathcal{L}}{\partial \theta_l^A}$ and $\frac{\partial \mathcal{L}}{\partial \theta_l^B}$, respectively. Then the two parties update their respective model with the decrypted gradients.

- **Step 6**: Party A sends termination signals to B once the loss $\mathcal{L}$ converges . Otherwise, goes to step 1 to continue the training process.

Recently, there are a large number of works discussing the potential risks associated with indirect privacy leakage through gradients [Bonawitz et al., 2016, Hitaj et al., 2017, McSherry, 2017, Phong et al., 2018, Shokri and Shmatikov, 2015]. To prevent the two parties from knowing each other's gradients, A and B further mask their own gradient with an encrypted random value. A and B then exchange the encrypted masked gradients and loss and obtain the decrypted values. Here, the encryption step is to prevent a malicious third-party from eavesdropping the transmissions, while the masking step is to prevent A and B from knowing each other's exact gradient value.

6.3.3   THE FTL PREDICTION PROCESS

Once the FTL model has been trained, it can be used to provide predictions for unlabeled data in party B. The prediction process for each unlabeled data point involves the following steps.

- **Step 1**: Party B computes u_j^B with the trained neural network parameters Θ^B, and sends encrypted $[[u_j^B]]$ to party A.

- **Step 2**: Party A evaluates u_j^B and masks the result with a random value, and sends the encrypted and masked $[[\varphi(u_j^B) + m^A]]_B$ to B.

- **Step 3**: Party B decrypts $[[\varphi(u_j^B) + m^A]]_B$ and sends $\varphi(u_j^B) + m^A$ back to party A.

- **Step 4**: Party A obtains $\varphi(u_j^B)$ and the label y_j^B, and sends the label y_j^B to B.

Note that the only source of performance loss over the secure FTL process is second-order Taylor approximation of the final loss function, rather than at every nonlinear activation layer of the neural network [Hesamifard et al., 2017]. The computations inside the networks are unaffected. As demonstrated in Liu et al. [2019], the errors in loss and gradient calculations, as well as the loss in accuracy by adopting our approach are minimal. Therefore, the approach is scalable and flexible to changes in neural network structures.

6.3.4   SECURITY ANALYSIS

As demonstrated in Liu et al. [2019], both the FTL training process and the FTL prediction process are secure under our security definition (see Definition 6.1), provided that the underlying additively homomorphic encryption scheme is secure.

During training, raw data $\mathcal{D}_A$ and $\mathcal{D}_B$, as well as the local models Net^A and Net^B are never exposed and only the encrypted hidden representations are exchanged. In each iteration, the only non-encrypted values party A and party B receive are the gradients of the model parameters, which are aggregated from all variables and masked by random numbers. At the end of the training process, each party (A or B) remains oblivious to the data structure of the other party and each obtains model parameters associated only with its own features. At inference time, the two parties need to collaborate in order to compute the prediction results.

Note the protocol does not deal with a malicious party. If party A fakes its inputs and submits only one non-zero input, it may be able tell the value of u_i^B at the position of that input. It still cannot tell x_i^B or Θ_B, and neither party will be able to obtain correct results.

6.3.5   SECRET SHARING-BASED FTL

Homomorphic encryption techniques are capable of providing a high level of security for the information or knowledge shared among parties, thereby protecting the privacy of data and models belonging to each party. However, homomorphic encryption techniques typically need extensive computational resources and massive parallelization to scale, which make them impractical in many applications that require real-time throughput.

An alternative secure protocol to homomorphic encryption is secret sharing. The biggest advantages of the secret sharing approach include (i) there is no accuracy loss, and (ii) computation is much more efficient than homomorphic encryption approach. The drawback of the secret sharing approach is that one has to offline generate and store many triplets before online computation.

To facilitate the description of secret sharing-based FTL algorithm, we rewrite Equations (6.6), (6.7), and (6.8) as follows:

$$\mathcal{L} = \mathcal{L}_A + \mathcal{L}_B + \mathcal{L}_{AB} \tag{6.9}$$

$$\frac{\partial \mathcal{L}}{\partial \theta_\ell^B} = \left(\frac{\partial \mathcal{L}}{\partial \theta_\ell^B} \right)_B + \left(\frac{\partial \mathcal{L}}{\partial \theta_\ell^B} \right)_{AB} \tag{6.10}$$

$$\frac{\partial \mathcal{L}}{\partial \theta_i^A} = \left(\frac{\partial \mathcal{L}}{\partial \theta_i^A} \right)_A + \left(\frac{\partial \mathcal{L}}{\partial \theta_i^A} \right)_{AB}, \tag{6.11}$$

where $\mathcal{L}_A$ and $(\frac{\partial \mathcal{L}}{\partial \theta_i^A})_A$ are computed solely by party A, and $\mathcal{L}_B$ and $(\frac{\partial \mathcal{L}}{\partial \theta_\ell^B})_B$ are computed solely by party B. $\mathcal{L}_{AB}$, $(\frac{\partial \mathcal{L}}{\partial \theta_\ell^B})_{AB}$ and $(\frac{\partial \mathcal{L}}{\partial \theta_i^A})_{AB}$ are computed collaboratively by A and B through secret sharing scheme.

The whole process of computing (6.9), (6.10), and (6.11) can be performed securely through secret sharing with the help of Beaver's triples. The secret sharing-based FTL training process is summarized in the following steps.

- **Step 1**: Party A and party B initialize and run their independent neural networks Net^A and Net^B locally to obtain hidden representations u_i^A and u_i^B.

- **Step 2**: Party A and party B collaboratively computes $\mathcal{L}_{AB}$ through secret sharing. Party A computes $\mathcal{L}_A$ and sends it to party B. Party B computes $\mathcal{L}_B$ and sends it to party A.

- **Step 3**: Party A and party B individually reconstruct loss $\mathcal{L}$ via Equation (6.9).

- **Step 4**: Party A and party B collaboratively computes $(\frac{\partial \mathcal{L}}{\partial \theta_\ell^A})_{AB}$ and $(\frac{\partial \mathcal{L}}{\partial \theta_\ell^B})_{AB}$ through secret sharing.

- **Step 5**: Party A computes its gradients via $\frac{\partial \mathcal{L}}{\partial \theta_\ell^A} = (\frac{\partial \mathcal{L}}{\partial \theta_l^A})_A + (\frac{\partial \mathcal{L}}{\partial \theta_\ell^A})_{AB}$ and updates its local model θ_ℓ^A. While at the same time, party B computes its gradients via $\frac{\partial \mathcal{L}}{\partial \theta_\ell^B} = (\frac{\partial \mathcal{L}}{\partial \theta_l^B})_B + (\frac{\partial \mathcal{L}}{\partial \theta_\ell^B})_{AB}$ and updates it local model θ_ℓ^B;

- **Step 6**: Party A sends termination signals to B once the loss $\mathcal{L}$ converges. Otherwise, goes to step 1 to continue the training process.

After training is completed, we proceed to the prediction phase. At the high level, the prediction process is quite simple. It involves the following two steps.

- **Step 1**: Party A and party B run their trained neural networks Net^A and Net^B locally to obtain hidden representations u_i^A and u_i^B.

- **Step 2**: based on u_i^A and u_i^B, party A and party B collaboratively reconstruct $\varphi(u_j^B)$ through secret sharing and calculate the label y_j^B.

Note that in both training and prediction processes, the only information that any party receives regarding any private value of the other party is only a share of that private value based on the secret sharing scheme. Therefore, no party is able to learn any information about the private values it is not supposed to learn.

6.4   CHALLENGES AND OUTLOOK

Traditional transfer learning is typically conducted in sequential or centralized way. Sequential transfer learning [Ruder, 2019] means that transfer knowledge is first learned on source task and then applied to target domain to improve the performance of the target model. Sequential transfer learning is ubiquitous and effective in computer vision where it is typically practiced in the form of pretrained model on large image datasets such as ImageNet [Bagdasaryan et al., 2009]. It is also commonly used in natural language processing to encode language units (e.g., word, sentence or document) in the form of distributed representations. The centralized transfer learning indicates that the models and data involved in transfer learning are located in one place. Thus, traditional transfer learning is not applicable in many practical applications where data is scattered among multiple parties and its privacy is a major concern. FTL is a feasible and promising solution to address those issues.

Research work on incorporating transfer learning into federated learning framework is fast-growing. However, for practical applications, FTL still faces many challenges. We list three of them as follows.

- We need to develop schemes to learn the transferable knowledge in a way that it can well capture the invariant between participants. Different from sequential and centralized transfer learning where the transfer knowledge is typically represented in one universal pre-trained model, transfer knowledge in FTL is distributed among local models. Each participant has total control in designing and training its local model. A balance should be achieved between autonomy and generalization performance of the FTL models.

- We need to determine how to learn a representation of transfer knowledge in a distributed environment while preserving the privacy of the shared representation of all participants. Under the federated learning framework, transfer knowledge is not only learned in a distributed manner, but also is typically not allowed to be exposed to any participant. Thus, we need to figure out precisely what each participant contributes to the shared representation in the federation and consider how to preserve the privacy of the shared representation.

- We need to design efficient secure protocols that can be employed in federated transfer learning. FTL usually requires closer interactions among participants in terms of communication frequency and the size of transferred data. Careful consideration should be taken when designing or choosing secure protocols in order to achieve a balance between security and overhead.

There are certainly many other challenges that are waiting for researchers and engineers to address. We envision that with the high practical value brought by FTL, more and more institutes and enterprises would invest resources and efforts into the research and implementation of FTL.

CHAPTER 7

Incentive Mechanism Design for Federated Learning

In federated learning, motivating data owners to continue participating in a data federation is an important challenge. The key to achieving this objective is to device an incentive scheme that shares the profit generated by a federation with participants in a fair and just manner. Before this step can be achieved, a mechanism for evaluating the contribution toward the federated model by a given data owner must be established. Although there has yet to be published research on solving this problem, there has been a well-established line of work on using auction-based approaches to motivate sensors to commit more resources to improve data quality, which might shed light into solving this problem.

In this chapter, we provide an overview of the problem of evaluating data owners' contributions and highlight some reverse auction-based approaches which are promising to be further developed to help address the problem of evaluating data owner contributions in federated learning. Following this discussion, we introduce a framework for fairness-aware profit sharing based on such evaluation results—the Federated Learning Incentivizer (FLI) payoff-sharing scheme [Yu et al., 2020]. It provides a blueprint for further advances in eliciting high quality contributions from data owners to be applied in situations in which data owners need to receive delayed payments for their rewards since the federation must use the federated model to generate revenue first.

7.1 PAYING FOR CONTRIBUTIONS

For a federation, data owners' continued participation in the federated learning process (e.g., through sharing of encrypted model parameters) is key to its long-term success. The contributions by data owners to a federation are used to build a machine learning (ML) model which, in turn, can be used to generate revenues. The federation can share part of the revenue with data owners as incentives (Figure 7.1). The research question here is how to quantify the payoff for each data owner in a context-aware manner in order to achieve long-term sustainable operation.

7.1.1 PROFIT-SHARING GAMES

Similar research problems have also been studied under the topic of cost-sharing games. In general, there are three categories of widely used profit-sharing schemes.

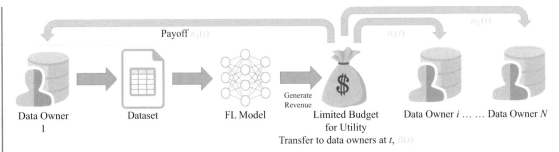

Figure 7.1: Transfer of utility from a data federation to its participants.

1. Egalitarian: any unit of utility produced by a data federation is divided equally among the data owners who helped produce it.

2. Marginal gain: the payoff of a data owner in a data federation is the utility that the team gained when she joined.

3. Marginal loss: the payoff of a data owner in a data federation is the utility that the team would lose if she were to leave.

In general, a participant i's payoff from a total budget $B(t)$ in a given round of profit-sharing t, denoted as $\hat{u}_i(t)$, is computed as:

$$\hat{u}_i(t) = \frac{u_i(t)}{\sum_{i=1}^{N} u_i(t)} B(t), \qquad (7.1)$$

where $u_i(t)$ is i's share of $B(t)$ among his peers computed following a given scheme.

Equal division is an example of egalitarian profit-sharing [Yang et al., 2017]. Under this scheme, the available profit-sharing budget, $B(t)$, at a given round t is equally divided among all N participants. Thus, a participant i's payoff is:

$$u_i(t) = \frac{1}{N}. \qquad (7.2)$$

Under the Individual profit-sharing scheme [Yang et al., 2017], each participant i's own contribution to the collective (assuming the collective only contains i), is used to determine his share of the profit, $u_i(t)$:

$$u_i(t) = v(\{i\}), \qquad (7.3)$$

where $v(X)$ is a function evaluating the utility of a collective X.

The Labor Union game [Gollapudi et al., 2017] profit-sharing scheme determines i's share of $B(t)$ based on his marginal contribution to the utility of the collective formed by his predecessors F (i.e., each participant's marginal contribution is computed based on the same sequence as they joined the collective):

$$u_i(t) = v(F \cup \{i\}) - v(F). \tag{7.4}$$

The Shapley game profit-sharing scheme [Augustine et al., 2015] is also a marginal contribution-based scheme. Unlike the Labor Union game, Shapley game aims to eliminate the effect of the participants joining the collective in different sequences in order to more fairly estimate their marginal contributions to the collective. Thus, it averages the marginal contribution for each i under all different permutations of the i joining the collective relative to other participants:

$$u_i(t) = \sum_{P \subseteq P_j \backslash \{i\}} \frac{|P|!(|P_j| - |P| - 1)!}{|P_j|} [v(P \cup \{i\}) - v(P)], \tag{7.5}$$

where a collective is divided into m parties $(P_1, P_2, \ldots, P_m)$.

The Fair-value game scheme [Gollapudi et al., 2017] is a marginal loss-based scheme. Under this scheme, i's share of the profit is determined by:

$$u_i(t) = v(F) - v(F \backslash \{i\}). \tag{7.6}$$

The sequence following which the participants leave a collective significantly affects his payoff.

7.1.2   REVERSE AUCTIONS

Other than profit-sharing game-based approaches, reverse auctions have also been used to develop incentive schemes that promote the quality of contributed data. There have been reverse auction schemes for sensor data [Singla and Krause, 2013], with the goal of finding the cheapest combination of sensors that provide data with the quality level. Such approaches are based on the assumption that the central entity knows what data is needed (e.g., geographical distribution). However, these approaches typically assume the data quality is independent of cost (since reverse auction requires identical items). They might also encourage arbitrage through submitting uninformative data just to collect rewards which is an unintended consequence.

Another way is to obtain data of a given quality is through posted rewards. This is a take-it-or-leave-it scheme. The federation could post a fixed reward to be paid for data owners who can contribute data with a certain quality level. The data owners can choose to participate in federated model training if their cost is lower than the reward; or not to participate if their cost is too higher than the reward. When effort from a data owner is required in order to meet the quality requirement, there are three existing categories of schemes for designing the rewards [Faltings and Radanovic, 2017]:

1. through output agreement [Dasgupta and Ghosh, 2013, Shnayder et al., 2016];

2. through information theoretic analysis [Kong and Schoenebeck, 2013]; and

3. through model improvement [Radanovic et al., 2016].

For gradient-based federated learning approaches, the gradient information can be regarded as a type of data. However, in these cases, output agreement-based rewards are hard to apply as mutual information requires a multi-task setting which is usually not present in such cases. Thus, among these three categories of schemes, model improvement is the most relevant way of designing rewards for federated learning. There are two emerging federated learning incentive schemes focused on model improvement.

Richardson et al. [2019] proposed a scheme which pays for marginal improvements brought about by model updates. The sum of improvements might result in overestimation of contribution. Thus, the proposed approach also includes a model for correcting the overestimation issue. This scheme ensures that payment is proportional to model quality improvement, which means the budget for achieving a target model quality level is predictable. It also ensures that data owners who submit model updates early receive a higher reward. This motivates them to participate even in early stages of the federated model training process.

Jia et al. [2019] is similar to Richardson et al. [2019], but computes a Shapley value to split reward among data owners. Such computations tend to be expensive. Instead, approximation by scaling factor adopted by Radanovic et al. [2016] can be much more computationally efficient. In addition, it does not address the issue that the same dataset can be contributed without extra cost to multiple federations.

These two schemes guarantee that uninformative data do not receive reward so as to discourage free-riders.

7.2   A FAIRNESS-AWARE PROFIT SHARING FRAMEWORK

The aforementioned schemes could be extended into situations in which the data owners are not paid upfront, but rather, have to wait for the federated model to generate revenue before receiving their rewards. In this section, we introduce a fairness-aware profit-sharing framework—FLI. It provides an architecture for incentive mechanism designers to consider such situations of delayed payment and incorporated fairness into such cases to sustain long-term participation by data owners.

7.2.1   MODELING CONTRIBUTION

The architecture of FLI is shown in Figure 7.2. We assume that the data federation follows synchronous mode of model training commonly adopted by federated learning [Bonawitz and Eichner et al., 2019] in which data owners share their model parameters in rounds. In round t, a data owner i can contribute his local model trained on a dataset to a federation. The federation is able to assess the contribution of i's data contribution to the federation following one of the profit-sharing schemes discussed in the previous section as the FLI baseline scheme.

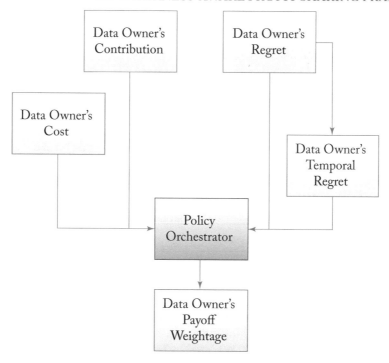

Figure 7.2: The overview of FLI.

To do so, a federation can run a sandbox simulation to estimate the effect of a data owner's contribution on model performance. The assessment outcome is recorded by a variable $q_i(t) \geq 0$, which denotes the expected marginal revenue the federated model can gain with i's latest contribution. The proposed incentive scheme is fully decoupled from how such a contribution score is produced. Thus, we do not focus on the exact mechanism by which $q_i(t)$ is produced, and assume the value is available to be used as an input for FLI.

7.2.2   MODELING COST

Let $c_i(t)$ be the cost for i to contribute $d_i(t)$ to the federation. There can be multiple ways to compute $c_i(t)$. Although it is possible to build computational models based on market research, a more practical solution is still auction-based self-report. A procurement auction [Mishra and Veeramani, 2007] can be used to estimate the cost when $c_i(t)$ is privately known. Specifically, the federation can ask each data owner to request a payment for the data contribution, and then select which data owner shall be allowed to join the federation.

In this case, the delayed payment scheme can be separated from the procurement auction where $c_i(t)$ can be interpreted as the payment to data owner i determined by the auction. This way, a clear separation of concern between the auction stage and the proposed incentive scheme can be achieved. Here, since we focuses on developing the framework of incentive design for federated learning, we leave the topic of computing $c_i(t)$ to be treated in another work, and assume that this value is available here.

7.2.3   MODELING REGRET

For each data owner i, the federation keeps track of the payoff gained from contributing data to the federation over time. As this value represents the difference between what the data owner has received so far and what he is supposed to receive, we refer to this term as *regret*, $Y_i(t)$. The dynamics of $Y_i(t)$ can be regarded as a queueing system:

$$Y_i(t+1) \triangleq \max\left[Y_i(t) + c_i(t) - u_i(t), 0\right], \tag{7.7}$$

where $u_i(t)$ is the payoff to be transferred to i by the federation. A large value of $Y_i(t)$ indicates that i has not been adequately compensated.

7.2.4   MODELING TEMPORAL REGRET

In some cases, the cost $c_i(t)$ may be too large to be fully covered by a single payment of $u_i(t)$ due to budget limitation in the federation. In such cases, the federation needs to compute instalments to be paid out to the data owners in multiple rounds. Their share of the current payout budget, $B(t)$, depends on their regret as well as how long they have been waiting to receive the full payoff.

For this purpose, we complement Equation (7.7) with a *temporal queue*, $Q_i(t)$, with queueing dynamics defined as:

$$Q_i(t+1) \triangleq \max\left[Q_i(t) + \lambda_i(t) - u_i(t), 0\right], \tag{7.8}$$

where $\lambda_i(t)$ is an indicator function:

$$\lambda_i(t) = \begin{cases} \hat{c}_i, & \text{if } Y_i(t) > 0 \\ 0, & \text{otherwise.} \end{cases} \tag{7.9}$$

This formulation means that as long as $Y_i(t)$ is not empty, the temporal queue, $Q_i(t)$, will increase. The increment is based on i's average cost of data contribution to the federation, $\hat{c}_i$, through past experience. Both queues decrease by the same amount when the federation pays i. The profit-sharing approach can ensure that data owners are compensated not only for their data contributions, but also for waiting to receive the full payoff, thereby making it "worth their while" to attract them to the federation.

7.2.5   THE POLICY ORCHESTRATOR

In order to encourage data owners to continue participating in the federation, the federation needs to ensure that the data owners are treated fairly based on their individual contribution. Here, we define three fairness criteria that are important to the long-term sustainable operation of a federation.

1. **Contribution Fairness**: a data owner i's payoff shall be positively correlated to his contribution to the federation $q_i(t)$.

2. **Regret Distribution Fairness**: the difference of the regret and the temporal regret among data owners shall be as small as possible.

3. **Expectation Fairness**: the fluctuation of data owners' regret and temporal regret values over time shall be as small as possible.

In order to satisfy all the three fairness criteria, the federation shall maximize a "value-minus-regret drift" objective function over time. The collective utility derived from data owners' contributions is related to two factors: (1) the contribution to the federation by a data owner i ($q_i(t)$) and (2) the payoff that i receives from the federation for the contribution ($u_i(t)$). It is fair that a data owner who make significant contribution to the federation shall receive high payoff. Thus, we have:

$$U = \frac{1}{T} \sum_{t=0}^{T-1} \sum_{i=1}^{N} \{q_i(t)u_i(t)\}. \tag{7.10}$$

Maximizing U satisfies **Fairness criterion (1)**.

Since $Y_i(0) = 0$ for all i, if we consistently strive to minimize the variation in $Y_i(t)$ over time, the regret must not grow unbounded to drive data owners away. Based on recommendations from the Belmont Report [1978], the federation needs to jointly consider the magnitude and distribution of regret among data owners and over time in order to treat them fairly [Yu et al., 2018]. l_2-norm can capture simultaneously the magnitudes of the regret values and the distribution of regret among data owners. A large l_2-norm value means there are many data owners with none-zero regrets, and/or there are a few data owners with very large regret [Yu et al., 2015, 2016, 2019]. Both shall be minimized.

Based on the l_2-norm technique, we formulate the Lyapunov function [Neely, 2010] of FLI as:

$$L(t) = \frac{1}{2} \sum_{i=1}^{N} \left[Y_i^2(t) + Q_i^2(t) \right]. \tag{7.11}$$

For simplicity of derivation later, we omit the $\sqrt{\cdot}$ operator in the standard l_2-norm calculation and multiply the whole term with $\frac{1}{2}$. These changes do not alter the desirable properties of l_2-norm for our formulation.

The drift in data owners' regret over time is:

$$
\begin{aligned}
\triangle &= \frac{1}{T} \sum_{t=0}^{T-1} [L(t+1) - L(t)] \\
&= \frac{1}{T} \sum_{t=0}^{T-1} \sum_{i=1}^{N} \left[\frac{1}{2} Y_i^2(t+1) - \frac{1}{2} Y_i^2(t) + \frac{1}{2} Q_i^2(t+1) - \frac{1}{2} Q_i^2(t) \right] \\
&\leqslant \frac{1}{T} \sum_{t=0}^{T-1} \sum_{i=1}^{N} \Big[Y_i(t)c_i(t) - Y_i(t)u_i(t) + \frac{1}{2}c_i^2(t) - c_i(t)u_i(t) + \frac{1}{2}u_i^2(t) + Q_i(t)\lambda_i(t) \\
&\qquad\qquad\qquad\qquad - Q_i(t)u_i(t) + \frac{1}{2}\lambda_i^2(t) - \lambda_i(t)u_i(t) + \frac{1}{2}u_i^2(t) \Big].
\end{aligned}
\tag{7.12}
$$

Since $u_i(t)$ it the control variable here, we extract only terms containing it from Equation (7.13):

$$
\triangle \leqslant \frac{1}{T} \sum_{t=0}^{T-1} \sum_{i=1}^{N} \left\{ u_i^2(t) - u_i(t)[Y_i(t) + c_i(t) + Q_i(t) + \lambda_i(t)] \right\}.
\tag{7.13}
$$

The regret drift variable $\triangle$ jointly captures the distribution of regret (both $Y_i(t)$ and $Q_i(t)$) among data owners, as well as the fluctuation of regret over time. Minimizing $\triangle$ satisfies **Fairness criteria (2) and (3)**.

By jointly considering collective utility and the distribution of regret, the overall objective function for a given federation can be defined as "maximizing collective utility while minimizing inequality among data owners' regret and waiting time":

$$
\omega U - \triangle
\tag{7.14}
$$

which shall be maximized. Here, ω is a regularization term for a federation to control the trade-off between the two objectives. Thus, the objective function of a federation is:

Maximize:

$$
\frac{1}{T} \sum_{t=0}^{T-1} \sum_{i=1}^{N} \left\{ u_i(t) [\omega q_i(t) + Y_i(t) + c_i(t) + Q_i(t) + \lambda_i(t)] - u_i^2(t) \right\}
\tag{7.15}
$$

Subject to:

$$
\sum_{i=1}^{N} \hat{u}_i(t) \leqslant B(t), \forall t
\tag{7.16}
$$

$$
\hat{u}_i(t) \geqslant 0, \forall i, t,
\tag{7.17}
$$

where $\hat{u}_i(t) \leqslant u_i(t)$ denotes the actual instalment payout from the federation to a data owner i in round t, which will be derived in the following section.

7.2.6 COMPUTING PAYOFF WEIGHTAGE

In order to optimize Equation (7.15), we set its first derivative to 0 and solve for $u_i(t)$:

$$\frac{d}{du_i(t)}[\omega U - \triangle] = 0. \tag{7.18}$$

Solving the above equation yields:

$$u_i(t) = \frac{1}{2}[\omega q_i(t) + Y_i(t) + c_i(t) + Q_i(t) + \lambda_i(t)]. \tag{7.19}$$

The second derivative of Equation (7.15) is:

$$\frac{d^2}{du_i^2(t)}[\omega U - \triangle] = -1 < 0 \tag{7.20}$$

indicating that the solution maximizes the objective function.

For contributing $d_i(t)$ amount of data of quality $q_i(t)$ at round t, the data owner i shall receive a total compensation of $u_i(t) = \frac{1}{2}[\omega q_i(t) + Y_i(t) + c_i(t) + Q_i(t) + \lambda_i(t)]$. The federation may need to pay out this in instalments over a period of time if not enough budget, $B(t)$, is available to pay all data owners fully at round t. To share $B(t)$ among the data owners, the computed $u_i(t)$ values are used as weights to divide the budget $B(t)$. The actual payout instalment to i at t, $\hat{u}_i(t)$, is:

$$\hat{u}_i(t) = \frac{u_i(t)}{\sum_{i=1}^{N} u_i(t)} B(t). \tag{7.21}$$

The FLI payoff-sharing scheme is summarized in Algorithm 7.1. It accounts for both the magnitude and the temporal aspects of participating in a federation. Data owners who has contributed a large set of high quality data, and who has not been fully compensated for a long time will enjoy a higher share of subsequent revenues generated by the federation.

The computational time complexity of Algorithm 7.1 is $O(N)$. Once $Y_i(t)$ and $Q_i(t)$ both reach 0 with no new cost incurred by i, $u_i(t) = \omega q_i(t)$. From then on, i will share future payoffs based on his contribution to the federation assessed using one of the baseline methods (e.g., the Shapley game payoff-sharing scheme). The proposed scheme prioritizes compensating the data owners with non-zero regret while taking into account their contributions to the federation.

7.3 DISCUSSIONS

In this chapter, we reviewed existing literature on profit-sharing games and reverse auctions which can be used to develop incentive mechanisms for federated learning, and highlighted some recent developments that leveraged some aspects of these related works to encourage early submission of high-quality contributions to the federation. Following this, we further proposed

---

Algorithm 7.1 Federated Learning Incentivizer (FLI)

Input: ω and $B(t)$ set by the system administrator; $Y_i(t)$ from all data owners at round t (with $Y_i(t) = 0$ for any i who just joined the federation); and $Q_i(t)$ from all data owners at round t (with $Q_i(t) = 0$ for any i who just joined the federation).

1: Initialize $S(t) \leftarrow 0$; //to hold the sum of all $u_i(t)$ values.
2: **for** $i = 1$ to N **do**
3:     **if** $d_i(t) > 0$ **then**
4:         Compute $c_i(t)$;
5:         Compute $q_i(t)$;
6:     **else**
7:         $c_i(t) = 0$;
8:     **end if**
9:     $u_i(t) \leftarrow \frac{1}{2}[\omega q_i(t) + Y_i(t) + c_i(t) + Q_i(t) + \lambda_i(t)]$;
10:    $S(t) \leftarrow S(t) + u_i(t)$;
11: **end for**
12: **for** $i = 1$ to N **do**
13:    $\hat{u}_i(t) \leftarrow \frac{u_i(t)}{S(t)} B(t)$
14:    $Y_i(t + 1) \leftarrow \max[0, Y_i(t) + c_i(t) - \hat{u}_i(t)]$;
15:    $Q_i(t + 1) \leftarrow \max[0, Q_i(t) + \lambda_i(t) - \hat{u}_i(t)]$;
16: **end for**
17: **return** $\{\hat{u}_1(t), \hat{u}_2(t), ..., \hat{u}_N(t)\}$

---

a architectural framework that allows the consideration of fairness to be incorporated into determining the priority for data owners to receive delayed payouts since the nature of federated learning business model is that the model must be built first before revenues can be generated to pay for incentives. It provides a way for a human controller to easily influence the weights of various factors being considered.

Much work still remains before the proposed mechanism can be operationalized. One of the most challenging task is to estimate data owners' cost incurred for joining the federation. Although it is possible to build computational models based on market research, a more likely solution is still auction-based self-report. A procurement auction can be used to estimate the cost when $c_i(t)$ is privately known. Specifically, the federation can ask each data owner to ask a payment for the data contribution, and then select which data owner shall be allowed to join the federation. In this case, the delayed payment scheme can be separated from the procurement auction where $c_i(t)$ can be interpreted as the payment to data owner i determined by the auction. This way, a clear separation of concern between the auction stage and the proposed payment scheme can be achieved.

Another challenge is how to estimate the contribution to the federation by a data owner i. A federation can run a sandbox simulation to estimate the effect of a data owner's contribution on model performance. A well-designed sandbox should be able to simulate the change in revenue as a result of a data owner's contribution. In this way, the mechanism is fully decoupled from how such a contribution score is produced.

CHAPTER 8

Federated Learning for Vision, Language, and Recommendation

In this chapter, we discuss the existing works of applying federated learning in computer vision, natural language processing, and recommender system, for enabling privacy-preserving AI applications.

8.1 FEDERATED LEARNING FOR COMPUTER VISION

Computer vision (CV) is the science of teaching machines to learn knowledge from images. It is a simulation of biological vision using computers and related equipment to learn 3D information of the corresponding scene from collected images and videos. In other words, we configure computers with "eyes" (cameras) and "brain" (algorithms), so that computers can perceive the world. Essentially, the core of CV is the study of how to organize the information on input images, detect objects and scenes, and then interpret the content of these images. From the perspective of solving a specific task, the research of CV can be classified into several categories, including Object Detection, Semantic Segmentation, Motion and Tracking, 3D Reconstruction, Visual Question and Answering, Action Recognition, and so on.

Ever since CV made its public debut in the 1980s, it has developed rapidly, transforming from shallow models combined with hand-crafted features (e.g., Histogram of Gradient (HOG) and Scale-Invariant Feature Transform (SIFT)) to end-to-end deep neural network (DNN) models. The traditional solutions of CV mostly follow the process of image pre-processing, feature extraction, training the model, and outputting the results. In deep learning (DL), a computer vision problem can be solved in an end-to-end manner that requires only the input of raw data and leaves other intricate engineering works to the machine.

8.1.1 FEDERATED CV

While computer vision has achieved unprecedented progress recently and it is leading the revolution on AI, this remarkable achievement is heavily driven by the availability of massive amount of image data. Typically, the most successful computer vision applications on the market are developed by organizations that have resources or user bases to collect sufficient data.

This resource-centric developing mode on DL, on the one hand, has boosted researches and developments on AI. On the other hand, it hinders the utilization of AI technologies in the massive majority of small companies, which normally only have limited data resources. One possible way to acquire data is through data sharing. However, for reasons such as data privacy, regulatory risks, and lack of incentives, companies are not willing to share their data with each other directly.

For instance, in the field of security, object detection techniques are commonly applied to detect abandoned and suspicious objects (Figure 8.1). Nevertheless, image data on those objects are imbalanced, and they are typically collected and labeled by different companies with different business goals. Due to data privacy and regulation concerns, those companies are not willing to share their data. On the other hand, they have strong motivation to build powerful object detection models that can improve their business margins.

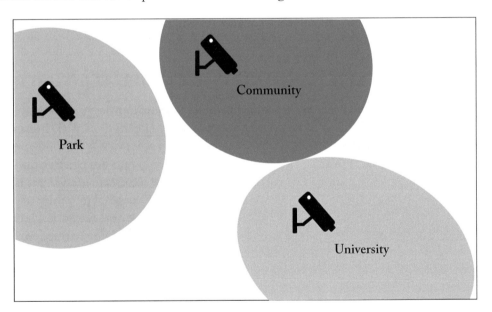

Figure 8.1: Illustration of federated object detection.

Federated learning can address these issues simultaneously. It enables multiple companies to collaboratively build a shared object detection model by exploiting all of their data without compromising data privacy. In addition, federated learning enables online feedback loop and model updates such that trained models can provide immediate response to customer's requests without delay. The workflow of federated object detection algorithm is illustrated in Figure 8.2 and the detailed steps are listed as follows.

- **Step 1**: Each participating company (i.e., data owner) downloads the current shared object detection model (e.g., YOLO [Redmon et al., 2016]) from the server.

- **Step 2**: Each company fine-tunes the model with locally annotated data.

- **Step 3**: Each company uploads parameters of the fine-tuned model to the server through secure protocol.

- **Step 4**: The server aggregates model parameters of all participants and updates the shared object detection model.

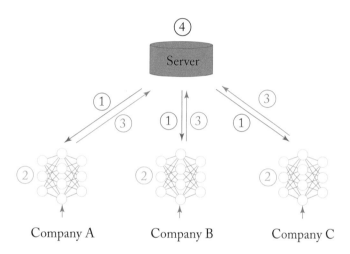

Figure 8.2: Federated learning workflow.

The federated object detection algorithm loops those steps until convergence. Then, local object detection models are deployed to work. Note that the whole training and deploying process can be conducted in a continuous manner as new annotated data constantly added in.

As illustrated in Figure 8.1, during the detection phase, cameras installed at population-intensive areas such as parks, shopping malls, and universities, use their locally deployed object detection models to detect suspicious objects. Local object detection models are constantly getting updated and deployed through the continuous federated learning procedure.

8.1.2   RELATED WORKS

Computer vision technologies are popular in the field of medical care for diagnosing and preventing diseases. Medical data are often stored at different institutions and are too sensitive to be directly shared due to privacy and legal concerns. To address this critical issue, federated learning presents a viable way for engineers and researchers. Sheller et al. [2018] introduced the first use of federated learning for multi-institutional collaboration, constructing a shared model without sharing patient data. Although to some extent, the proposed federated learning approach addresses the privacy issue of sharing medical images, it still requires a trusted entity for collecting and aggregating updates of local models. This centralized mode might introduce backdoors for malicious participants and have the risk of a single point of failure.

To address this issue, BrainTorrent [Roy et al., 2019] proposed a novel federate learning framework that does not depend on a central server. Different from centralized FL, it randomly chooses a client as the temporary server that sends out a ping request to the rest of clients for checking their versions and aggregating model parameters of a new version. For the analysis of neuroimaging data, such as magnetic resonance imaging (MRI) scans, Silva et al. [2018] came up with an end-to-end FL framework for securely accessing and meta-analyzing any biomedical data without sharing individual information. In addition, this framework adopts schemes analysis through the Alternating Direction Method of Multipliers (ADMM) for the potential bottlenecks of gradient-based optimization, which could cut down the number of iterations. This framework includes three main steps: (1) data standardization, a data preprocessing step to enhance the stability of the analysis and ease the comparison across features; (2) correction of confounding factors that have biasing effects on the data; and (3) multivariate analysis of variability, transforming the high-dimensional features into low dimensional representations via federated principal component analysis (PCA).

As DL currently is dominant in computer vision, models (i.e., deep CNN models) developed to tackle sophisticated imaging tasks are typically complex and massive. For instance, it could take hundreds of megabytes to store an object detection model. Naturally, it can also take a significant amount of time to train a large CNN model. To boost the training process, pre-trained models are commonly used for speeding up convergence.

However, pre-trained ML models are not compatible with existing federated learning scenarios, where local and global models are learned together. Yurochkin et al. [2019] investigated this problem and developed a probabilistic federated learning framework that can aggregate pre-trained neural network models. More specifically, the proposed federated learning idea is to match the parameters of trained local models across clients to construct a global model. The matching is governed by the posterior of a Beta–Bernoulli process (BBP) [Thibaux and Jordan, 2007] that allows the local parameters to be combined into a federated global model without knowing additional data or knowledge of learning algorithms used to learn the pre-trained models. The biggest benefit brought by this novel federated learning approach is that it decouples the training of local models from the federating of a global model. This decoupling makes model pre-training strategy compatible with federate learning scenarios and allows it to be agnostic about local learning algorithms.

8.1.3   CHALLENGES AND OUTLOOK

Although a series of novel research has been conducted on applying federated learning to solve computer vision problems, the large size of deep CNN models hinders federated learning from being adopted in practical applications in mobile or embedded devices. Federated learning brings model training to client devices. On one hand, this obviates the need for centralizing users' private data. On the other hand, it imposes a great challenge on client devices that typically have limited computing power. With challenges come opportunities, in addition to pushing mobile devices makers such as Apple, Huawei, and Xiaomi to develop specialized hardware tailored for training deep neural networks, the soaring demands on smart on-device applications may trigger advancements in model compression techniques such as parameter pruning, low-rank factorization, knowledge distillation, and many others, to significantly reduce the size of convolutional neural network models, thereby saving computing resources and communication cost.

Probably one of the most promising and challenging applications in which federated learning can really shine is the CV-powered autonomous driving system built based on heterogeneous data scattered among various types of devices. Autonomous car makers have long been searching for stable and robust approaches to ensure the safety of drivers in each and every situation. Nonetheless, as you may have heard of that only a few stickers can trick deep neural networks into miss-classifying traffic signs, you might be concerned about this kind of limitations of neural networks. However, we argue that enriching the information sources based on which the car makes decisions would be more helpful for the safety of drivers. When we make a judgment while driving, we use all of our physical senses, including sight, touch, hearing, and even smell. Analogously, an autonomous car should maneuver by negotiating with all sorts of devices (such as cameras, radar, and lidar) that are not only from itself but also from surrounding cars, and even monitors on the road.

Federated learning can power autonomous driving systems by uniting all kinds of devices to collaboratively build shared and personalized models. These models are well informed and can make smart and comprehensive decisions. However, this cannot be achieved within one day. Other than intelligent algorithms that can effectively learn patterns from distributed and heterogeneous data (e.g., images, voice signals, and other numerical data), advanced communication protocols should be developed to support real-time interaction among various devices and efficient secure protocols are also required to guarantee the privacy and security of drivers' personal data.

Therefore, while federated learning is a promising approach to achieve more intelligent and securer AI applications, a significant amount of resources and efforts should be devoted to making it happen.

8.2   FEDERATED LEARNING FOR NLP

Natural language processing (NLP) models have achieved remarkable success for a wide range of tasks with the advances in deep neural networks (DNNs). Among these neural network models, Recurrent Neural Networks (RNNs) that are able to efficiently process temporal information in sequences have significantly improved the performance of language modeling. Popular variants of RNNs include Long Short-Term Memory (LSTM) [Hochreiter and Schmidhuber, 1997] and Gated Recurrent Unit (GRU) [Cho et al., 2014].

However, these approaches require training data generated from use by many users to be aggregated into a central storage. In real-world scenarios, the users' natural language data are sensitive and may include private content. Therefore, in order to protect the privacy of users while still be able to build powerful NLP models, federated learning techniques are leveraged.

8.2.1   FEDERATED NLP

One typical application of federal learning in NLP is to learn out-of-vocabulary (OOV) words based on frequently typed words from mobile device users [Chen et al., 2019]. OOV refers to words that are not included in the vocabulary of a user's mobile device. Words missing from the vocabulary cannot be predicted by keyboard suggestion, auto-corrected, or gesture-typed. To learn an OOV generation model from a single user's mobile device is impractical since each user's device normally only stores a vocabulary of a limited size. Collecting data of all users to train an OOV generation model is not feasible either since OOV words typically contain sensitive user contents. In this scenario, federated learning is especially useful since it can train a shared OOV generation model based on data of all mobile users without the need for transmitting sensitive contents to centralized servers.

Any sequence model, such as LSTM, GRU, and WaveNet [van den Oord et al., 2016], can be used to learn OOV words. The federated OOV algorithm conducts a similar workflow as illustrated in Section 8.2. It iteratively performs the following steps to train the shared OOV generation model until convergence.

- **Step 1**: Each mobile device downloads the shared model from the server.

- **Step 2**: Each mobile device trains the shared model based on user-typed content.

- **Step 3**: Each mobile device summarizes the changes as a small focused update and uploads this update to the server through secure protocol.

- **Step 4**: The server gathers the updates from mobile devices, aggregates these updates, and improves the shared model with the aggregated updates.

During the federated learning process, the OOV generation model located in each user's mobile device is constantly getting updated, while the training data remains on the device. As a result, each mobile device ends up with a powerful OOV generation model. As illustrated in Figure 8.3, equipped with the federated OOV generation model encoding input information of all users, each user's mobile device can provide the user with rich and diverse query suggestions. Note that the user has total control in deciding when to join or leave the federated learning. Therefore, the server should have an analytic mechanism to monitor devices' health statistics such as how many devices join or leave the federated learning process per training round. Interested readers can refer to Bonawitz and Eichner et al. [2019] for more details on designing a scalable federated learning system.

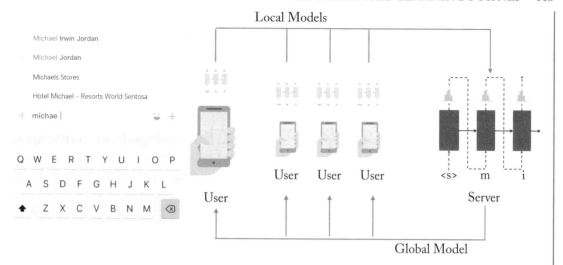

Figure 8.3: Federated learning for OOV generation.

8.2.2   RELATED WORKS

In addition to learning federated OOV generation models, federated learning can be applied to on-device wake-word detection. An example of wake-word detection is to say "Hello Siri!" to an iPhone to wake up the speech recognition and language understanding module. Wake-word detection is a key component that enables voice-based user interaction on smart devices. As they are always-on in nature, wake-word detection applications must only require a highly limited power budget and typically run on micro-controllers with limited memory and computing resources. Besides, they must maintain behavioral consistency in a variety of complex situations and show robustness against background noise. In addition, they are expected to have high recall on command capture and low false alarm rate.

Snips [Leroy et al., 2019] recently published a work on investigating the use of federated learning to train a resource-constrained wake-word detector based on a crowdsourced dataset that simulates the real-world non-i.i.d, unbalanced, and highly distributed setting. This work optimizes the federated averaging algorithm through an adaptive averaging strategy inspired from Adam. Benefited from this optimization, the federated learning algorithm only costs 8MB per client for communication and converges within 100 communication rounds.

The attention mechanism has been widely used in in sequence-to-sequence natural language processing tasks such as neural language translation and image caption generation. A recent work [Ji et al., 2018] introduced the attention mechanism into the federated setting of mobile keyboard suggestion—the Attentive Federated Aggregation (FedAtt) approach. Different from traditional attention mechanisms applied to the data flow, the FedAtt takes the server parameters as a query and the client parameters as keys, and calculates the attention weight in each layer of the GRU neural network for each client. Then, it updates each layer of the server model through the aggregation of the attention-weighted parameters from the same layer of the clients' models. The biggest benefit brought by FedAtt is that the global server model is fine-tuned through fine-grained aggregation of client models.

8.2.3   CHALLENGES AND OUTLOOK

In the field of NLP, the dominant approach is supervised learning. In order to achieve decent performance on data that the model has not come across before, supervised learning requires a sufficiently large amount of labeled training data for every new scenario. It is extremely expensive in time and labor to annotate every single training data manually. Federated learning can address this issue to some extent by exploiting data from many participating parties. However, for scenarios where labeled data are extremely scarce, all available data (both labeled and unlabeled) should be utilized to train NLP models.

Federal learning combined with unsupervised learning, semi-supervised learning, or transfer learning is a promising research direction to address the data-scarcity problem. Particularly in the field of NLP, a large amount of data are unlabeled. How to effectively exploit those data is an interesting and challenging research topic. Many innovative researches have already been conducted in unfederated learning settings such as cross-lingual learning [Ruder et al., 2017], multi-task learning [Augenstein et al., 2018], and multi-source domain adaptation [Chen and Cardie, 2018]. Under the federated learning setting, effectively utilizing unlabeled data is even more challenging. For instance, we need to determine how to learn transferrable or disentangled representations from unlabeled data in a distributed environment while preserving participants' privacy. Also, we need to carefully design or choose secure protocols in order to achieve a balance between security and efficiency. There are certainly many other challenges and difficulties that we will encounter during the process of developing federated learning. Nonetheless, we envisage that the advances in NLP under federated setting would significantly expand the application of artificial intelligence to many more industries.

8.3   FEDERATED LEARNING FOR RECOMMENDATION SYSTEMS

Shopping is a necessity of our everyday life. We used to purchase stuff in brick-and-mortar stores and consult people we trust, such as friends, family or shop keepers. The Internet has revolutionized the way we shop. Online shopping has become extremely popular nowadays. With a simple click on the search button, thousands of related products will pop out instantaneously. During this process, whether we realize it or not, we are using recommendation systems (a.k.a. recommender systems). In fact, recommendation systems are ubiquitous. When we are looking for books on Amazon, searching for hotels on Expedia, or browsing photos on Instagram, we are taking advantage of, and at the same time making contributions to, the recommendation systems.

What is a recommendation system (RS)? Briefly speaking, a recommendation system is an information filtering tool that utilizes user profiles and habits of the whole community to present the most relevant items that a particular user might be interested in. An effective recommendation system has three main functions.

- Overcoming the information overload problem. With the explosive growth of information on the Internet, it is impossible for users to go through all the contents. Recommendation system can filter out low-valued information and, thus, save users' time.

- Providing personalized recommendations. Users who have highly specific preferences often have difficulties finding their favorite items. Recommendation system should help users better locate what they are truly interested in based on their tastes.

- Making rational use of resources. According to the long tail effect, the most popular items attract most of the attention while the less popular items, which are the majority of the offered products, are rarely visited. This is a huge waste of resources. Recommendation systems should balance popularity and utility, and give more exposure to these less popular items.

An effective recommendation system benefits both the platform and the company. Users are more likely to click or purchase items recommended based on their preferences and revisit websites that know their habits better. In all, recommendation systems have been playing a vital role in various information retrieval systems to boost business and facilitate decision making [Zhang et al., 2019].

However, there are still many unresolved problems in recommendation systems. Cold start and user data privacy are the two major issues. Federated learning is promising for solving these two problems simultaneously. Imagine that we are building a global model using data from multiple parties through federated learning. For the cold start problem, we could borrow knowledge from other parties that might have relevant information to help rate a new item or make a prediction for a new user. For the data privacy issue, user's private data are kept in the client device and only model updates are uploaded through secure protocols. Moreover, federated learning distributes the model learning process to the clients' ends, which greatly reduces the computational pressure on the central server.

8.3.1   RECOMMENDATION MODEL

Before diving into the federated recommendation system, we first introduce the current recommendation models. In general, recommendation models can be classified into four categories: (1) collaborative filtering, (2) content-based, (3) model-based, and (4) hybrid recommender system [Adomavicius and Tuzhilin, 2005].

1. Collaborative Filtering (CF) makes recommendations by modeling users' historical interactions including explicit feedback (e.g., previous ratings) and implicit feedback (e.g., browsing and purchase history). That is to say, CF recommends a new item to a user if similar users purchased the item or the user purchased similar items. However, in reality, each user often interacts with only a few items, making the user-item interaction matrix highly sparse. Low-rank factorization approaches, also called matrix factorization, have been proven to be effective to solve this sparsity problem [Zhou et al., 2008].

2. Content-based recommendation is based on comparisons between the description of an item and the profile of a user. The core idea is that if a user likes an item, he will probably also like similar items. In a content-based recommendation system, items are labeled by several keywords and a user's profile consists of keywords describing what types of items the user prefers. The model recommends products the description of which matches the user's profile through keyword alignment.

3. Model-based recommendation system uses ML and DL technologies to directly model user-item relationships. It has several advantages: (1) compared to the former two linear approaches, it is capable of modeling nonlinear relationships; (2) DL models are able to learn latent representations of heterogeneous information such as text, images, and audios, which produces better recommender models; and (3) DL models such as Recurrent Neural Networks (RNN) are able to process sequential data so that they are suitable in sequential pattern mining tasks such as next-item prediction.

4. The hybrid recommendation system refers to the model that integrates two or more types of recommendation strategies, which is usually considered more effective. A simple hybrid approach is to make collaborative and content-based predictions separately and then combine their results. Take movie recommendation as an example. The hybrid model makes recommendations for a user based on movie watching and search histories from similar users (collaborative-based filtering) as well as movies that are similar to the ones the user favors (content-based filtering).

8.3.2   FEDERATED RECOMMENDATION SYSTEM

In this section, we will use the federated collaborative filtering as an example to briefly describe how a federated recommendation system works. Interested readers may refer to Ammad-ud-din et al. [2019] for more details.

Suppose that an e-commerce company wants to train a collaborative filtering (CF) model for users to find desirable items based on both personal preference and popularity. Since users' raw data cannot be directly collected due to data privacy and security reasons, federated learning can be exploited to train the CF model. Typically, a CF model can be formulated as a combination of a user factor matrix consisting of multiple user factor vectors that each represents a user,

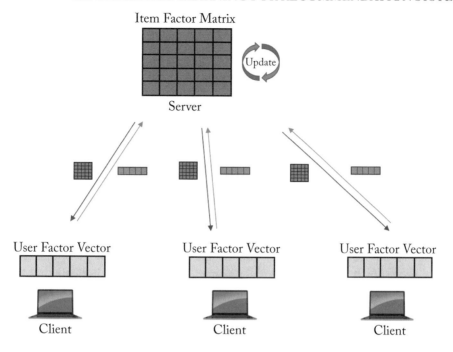

Figure 8.4: Federated collaborative filtering.

and an item factor matrix consisting of multiple item factor vectors that each represents a product item. The federate CF is to learn the two matrices collaboratively by all clients (illustrated in Figure 8.4), which involves the following five steps.

- **Step 1**: Each client (i.e., user's local device) downloads the global item factor matrix from the central server. This matrix can be either randomly initialized or pre-trained.

- **Step 2**: Each client gathers both explicit and implicit data. Explicit data includes user's feedback such as ratings and comments on products. Implicit data consists of user order history, cart list, browsing history, click history, search log, and so on.

- **Step 3**: Each client updates its local user factor vector using the local data and the global item factor matrix.

- **Step 4**: Each client computes the local update for the global item factor matrix using its local data and local user factor vector, and uploads the update to the central server via a secure protocol.

- **Step 5**: The central server updates the global item factor matrix based on the aggregation of local updates through federated weighting algorithms, e.g., federated averaging algorithm [McMahan et al., 2016b]. Then, the server sends the global item factor matrix to each client.

Above procedure is a general case of federated collaborative filtering. We can replace the CF model with more powerful models, e.g., deep FM model [Guo et al., 2017] to further improve the performance. In addition to personalized recommendation task, federated recommendation systems can also benefit companies by combining heterogeneous datasets, i.e., combining different feature dimensions from different parties to improve recommendation accuracy.

8.3.3   RELATED WORKS

Although federated learning is a new research area, researchers have made some progress on integrating federated learning with recommendation systems, showing promising results.

Kharitonov [2019] applied federated learning in an online learning setup, namely federated online learning to rank (FOLtR). He combines evolution strategy optimization with a privatization procedure based on differential privacy. The experiments indicate that FOLtR-ES performs close to RankingSVM and MSE baselines and is robust against privatization noise.

Chen et al. [2018] proposed a federated meta-learning framework for the recommendation that shares user information at the algorithm level. The system utilizes a high-level parameterized algorithm to train parameterized recommendation models (i.e., both the algorithm and the model are parameterized and need to be optimized). Moreover, the local model is user-specific and can be kept at a small scale to reduce resource consumption. The experiments show that the federated meta-learning recommendation models achieves the highest prediction accuracy compared to baselines and can be adapted quickly to new users in a few update steps.

Trienes et al. [2018] treated federated learning as a decentralized network, which is able to resolve issues such as large-scale user-surveillance and misuse of user data to manipulate elections. They investigated how to apply recommendation algorithms to decentralized social networks and implemented a collaborative filtering recommender and a topology-based recommender based on large unbiased samples collected from a federated social network named Mastodon. The experiments show that collaborative filtering approaches outperform a topology-based approach.

8.3.4   CHALLENGES AND OUTLOOK

We can see that researchers have carried some innovative work on combining federated learning and recommendation systems. There are still a lot of gaps in this field to be bridged. A general question is: what are needed to build practical privacy-preserving and secure recommender systems and how can we build them? It can be further divided into several detailed questions: (1) how do you achieve high accuracy and low communication cost while persevering data security and privacy; (2) which secure protocols should we choose; and (3) which recommendation algorithms are more suitable under federated learning?

Here are some possible research directions for future work. First, to what extent will incomplete data impact the performance of a recommendation system? In other words, what is the minimum amount of data we need to collect from the users in order to build an accurate recommendation system? Second, traditional recommenders utilize user social data, spatiotemporal data, and so on. However, it is still unclear which part of such data is more useful. Third, federated learning framework is very different from the traditional setting of recommendation systems. Therefore, which recommendation algorithm is more suitable or more robust under federated learning?

CHAPTER 9

Federated Reinforcement Learning

In this chapter, we introduce federated reinforcement learning (FRL), covering the basics of reinforcement learning, distributed reinforcement learning, horizontal FRL and vertical FRL, as well as application examples of FRL.

9.1   INTRODUCTION TO REINFORCEMENT LEARNING

Reinforcement Learning (RL) is a branch of machine learning (ML) which mainly deals with sequential decision-making [Sutton and Barto, 1998]. An RL problem usually consists of a dynamic environment and an agent (or agents) which interacts with the environment. The environment evolves once the agent selects an action based on the current state of the environment by presenting a reward for evaluating the performance of the agent. The agent seeks to achieve a goal in the environment by making sequential decisions. Traditional RL problems can be formulated as Markov Decision Processes (MDPs). The agent has to tackle a sequential decision-making problem to maximize a value function (i.e., the expected sum of discounted rewards, or reward expectations).

As shown in Figure 9.1, the agent observes the environment state, then selects an action based on the state. The agent is expected to receive a reward from the environment based on this selected action. In MDPs, the next state of the environment is dependent on the last state and the action selected by the agent. The action which results in the "highest expected reward" usually refers to the action that puts the agent in the state with the highest potential to gain rewards in the future. The agent moves in state-action-reward-state (SARS) cycles.

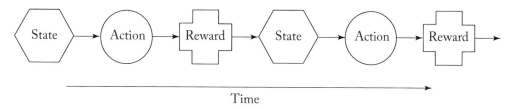

Figure 9.1: The state-action-reward-state cycle.

The difficulties of this problem lie in the following issues.

- An agent has limited knowledge about the optimal actions for a given state of the environment. Considering the one-round decision-making process in the RL problem with continuous action space, the agent has to deal with an optimization problem with continuous space, which may require a huge computation effort.

- The agent's actions can affect the future states of the environment, thereby affecting the options and opportunities available to the agent in the future. Therefore, dealing with sequential decision-making problems, each agent should not greedily choose actions even though these actions may obtain good rewards in a short term. That is, the agent has to trade-off between the current reward and the future reward expectations.

- Selecting optimal actions requires taking into account indirect, delayed consequences of the actions, and thus may require foresight or planning.

Other than the agent and the environment, one can identify four key sub-elements of an RL system: policy, reward signal, value function, and optionally, a model of the environment.

9.1.1   POLICY

A policy defines the agent's decisions to choose an action given a specified state. Roughly speaking, it is a mapping function (or conditional distribution) that takes the current state of the environment as input and returns the optimal (or sub-optimal) actions. A policy is the core of an RL agent as it determines the mapping from environment state to agent behavior, which can be deterministic or stochastic.

9.1.2   REWARD

A reward defines immediate feedback from the environment to the agent in an RL problem. At each time step, after an agent acts according to its current policy, the environment presents a reward that justifies the performance of the agent's actions. The agent's sole objective is to maximize the total reward over the long run. The reward signal is the primary basis for adjusting the policy.

9.1.3   VALUE FUNCTION

Whereas the reward signal feeds back to the agent what is good in an immediate sense, a value function predicts the expected future rewards the agent may accumulate starting from the current state. Roughly speaking, the value function is a way to measure the performance of an action under a given state, and the purpose of estimating values is to accumulate higher rewards. In traditional value-based RL methods, an action should be chosen based on the highest value rather than the highest reward, since the value function evaluates the reward expectations in the long run. Rewards are basically given directly by the environment, but values must be estimated and re-estimated from the sequence of observations an agent makes over its entire lifetime.

9.1.4   MODEL OF THE ENVIRONMENT

For some reinforcement learning systems, there can be a model of the environment. The model of the environment is a virtual model that mimics the behavior of the environment. For example, given a state-action pair, the model of the environment might predict the resultant next state and next reward and, with these predictions, possible future situations before they actually happen can be considered. Methods for solving reinforcement learning problems that use models are called model-based methods. However, most other algorithms assume that the model is not known, and they estimate the policy and value function by trial-and-error. These methods are known as model-free methods.

9.1.5   RL EXAMPLE

In the following, we present a detailed background example for RL description: the optimal control of the coal-fired boilers in the power plants. The coal-fired boiler systems play roles in transforming the energy firstly into steam heat and then into electricity, which is of a highly dynamic nature. The stochastic factors may come from the unpredictable changes in demands, the equipment conditions, and the calorific values of coal, etc. Figure 9.2 presents a basic framework for applying RL methods to optimal control of coal-fired boiler systems.

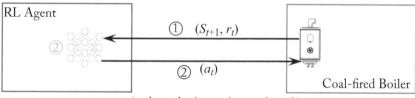

a_t = (coal_speed, primary_air, secondary_air)
S_t = (boiler_temp, gas_oxygen, steam_pressure)
r_t = power efficiency, temperature stability, etc.

Figure 9.2: RL framework for coal-fired boiler control.

As can be seen from Figure 9.2, in order to train an RL agent for the optimal control of a coal-fired boiler system, the following interactions have to be conducted.

1. *RL agent gets observations.* The RL agent can obtain the observations of the coal-fired boiler, including boiler temperatures, flue gas oxygen content, steam pressure, etc.

2. *RL agent takes actions.* Then, based on the learned knowledge of the RL agent, an action is presented to the control system of the coal-fired boiler. The action includes the speed of the coal conveying belt, primary air volume, secondary airflow, etc.

3. *Coal-fired boiler evolves.* Finally, the coal-fired boiler accepts the agent's actions and evolves into another condition.

9.2   REINFORCEMENT LEARNING ALGORITHMS

RL algorithms can be categorized according to the following key elements.

Model based vs. Model free: Model-based methods attempt to build a virtual model of the environment first and then act according to the best policy derived from the virtual model. For model-free methods, they assume the model of the environment cannot be built, and they estimate the policy and value function by trial-and-error.

Value based vs. Policy based: Value-based methods attempt to learn a value function and infer an optimal policy from it. Policy-based methods directly search in the space of the policy parameters to find an optimal policy.

Monte Carlo Update vs. Temporal Difference (TD) Update: Monte Carlo Update evaluates a policy using the accumulated reward over the entire episode. This is straightforward in implementation, but they require a large number of iterations for convergence and suffer from a large variance when estimating their value function. Instead of using the total accumulated reward, TD Update calculates a temporal error, which is the difference of the new estimate of the value function and the old estimate of the value function, to update the policy. This kind of update only needs the most recent iterations and reduces the variance. However, the bias increases during estimation as the global view of the whole episode is not considered.

On-policy vs. Off-policy: On-policy methods use the current policy to generate actions and update the current policy itself accordingly. Off-policy methods use a different exploratory policy to generate actions and the target policy is updated based on these actions.

Table 9.1 is a summary of some popular RL algorithms and their categories. Two TD algorithms that have been widely used to solve RL problems are State-Action-Reward-State-Action (SARSA) and Q-Learning.

SARSA is an On-policy, TD algorithm [Rummery and Niranjan, 1994]. It is on policy since it follows the same policy to find the next action. It tries to learn an action-value function instead of a value function. The policy evaluation step uses the temporal error for the action-value function, which is similar to the value function.

Q-Learning is an Off-policy, TD algorithm [Watkins and Dayan, 1992]. It is Off-policy since it selects the next action in a greedy fashion rather than follow the same policy. The Q-function is updated by using a policy that is directly greedy with respect to the current Q-function.

9.3   DISTRIBUTED REINFORCEMENT LEARNING

RL algorithms can be interpreted as playing a game many times and learning from a huge number of trials. This can be very time consuming if only one agent is involved to explore a huge

Table 9.1: The state-action-reward-state cycle

	Model-Free	Model-Based	Policy-Based	Value-Based	Monte Carlo Update	Temporal Difference Update	On-Policy	Off-Policy
Q-learning	✓	✓		✓		✓		✓
SARSA	✓	✓		✓		✓	✓	
Policy Gradients	✓	✓	✓		✓			
Actor-critic			✓	✓				
Monte-Carlo Learning					✓			
SARSA Lambda							✓	
Deep Q-Network								✓

state-action space. If we have multiple copies of the agent and the environment, the problem can be solved more efficiently in a distributed fashion. The distributed RL paradigm can be either Asynchronous or Synchronous.

9.3.1   ASYNCHRONOUS DISTRIBUTED REINFORCEMENT LEARNING

In the asynchronous scenario, multiple agents explore their own environments separately, and a global set of parameters is updated asynchronously. This allows a large number of actors to learn collaboratively. However, due to the delay of some agents, this algorithm may suffer from stale (old) gradients problem.

A3C Asynchronous Advantage Actor-Critic (A3C) is an algorithm proposed by Google DeepMind in 2016 [Mnih et al., 2016]. It creates up to 16 (or 32) copies of the agent and environment on a single CPU when learning Atari benchmark games. As the algorithm is highly paralleled, it is able to learn many of the Atari benchmark games very quickly on inexpensive CPU hardware.

General Reinforcement Learning Architecture (Gorila) [Nair et al., 2015] is another asynchronous framework for large scale distributed reinforcement learning. Multiple agents can be created and they are separated into different roles including actors and learners. Actors only generate experience by acting in the environment. The collection of experience is stored in a shared replay memory. Learners only train by sampling from the replay memory.

9.3.2   SYNCHRONOUS DISTRIBUTED REINFORCEMENT LEARNING

Sync-Opt Synchronous Stochastic Optimization (Sync-Opt) [Chen et al., 2017] attempts to solve the problem of slow, straggling agents which slow down synchronous RL learning. To avoid these agents, the training process only waits for a preset number of agents to return, and the slowest few agents are dropped.

Advantage Actor-Critic (A2C) [Clemente et al., 2017] is a modified version of the famous A3C. It works in the same way, with the exception that it synchronizes all agents between rounds. OpenAI claims in their blog post that synchronous A2C outperforms A3C [Mnih et al., 2016].

9.4   FEDERATED REINFORCEMENT LEARNING

DRL has many technical and non-technical issues during implementations. One of the most critical issues is how to prevent information-leakage and to preserve agent privacy during DRL. This concern leads to privacy-preserving versions of RL—Federated Reinforcement Learning (FRL). Here we categorize FRL researches into Horizontal Federated Reinforcement Learning (HFRL) and Vertical Federated Reinforcement Learning (VFRL).

This section illustrates the background of HFRL and VFRL. We emphasize on presenting the detailed backgrounds, problem settings and possible framework for both HRFL and VFRL using the background example in real-life industry systems presented in Section 9.1.5.

FRL Background

In RL researches, one of the commonly studied interests is to design feedback controllers for discrete- and continuous-time dynamical systems that combine features of adaptive control and optimal control. These feedback control problems include self-driving systems, autonomous helicopter control and optimal control of industrial systems, etc.

Note that Zhuo et al. [2019] presented several real-life FRL examples.

1. In manufacturing, factories produce different components. The decision policies are private and will not be shared with each other. On the other hand, building high-quality individual decision policies is often difficult due to limited businesses and lack of reward signals (for some agents). It is thus helpful for factories to learn decision policies cooperatively under the condition that private data are not given away.

2. Another example is building medical treatment policies for hospital patients. Patients may be treated in some hospitals without providing feedback about the treatments (i.e., no reward signal for these treatment decision policies). In addition, data records about patients are private and may not be shared among hospitals. It is thus necessary to learn treatment policies for hospitals through Federated DRL.

In the following chapters, for the sake of consistency, we will explain the detailed background, problem settings and framework of HFRL and VFRL based on coal-fired boiler systems.

Horizontal FRL

Parallel RL [Kretchmar, 2002, Grounds and Kudenko, 2005] has long been studied in the RL research community, in which multiple agents are assumed to perform the same task (with the same rewards corresponding to states and actions). The agents may carry out learning in different environments. Note that most parallel RL settings adopt the operations of transferring agents' experience or gradients. It is straightforward that such operations cannot be conducted when considering privacy-preserving issues. Therefore, it is natural to adopt HFRL for privacy-preserving issues. The HFRL community adopts these basic settings in parallel RL with the privacy-preserving objective as an extra constraint (for both the server and the agents). A basic framework for conducting HFRL is presented in Figure 9.3.

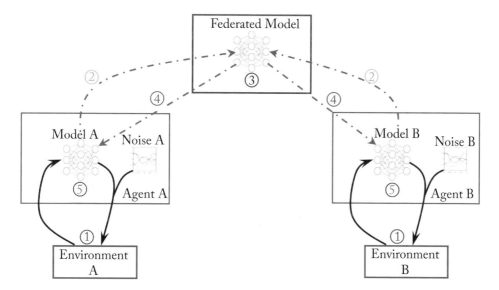

Figure 9.3: Example architecture of HFRL framework.

As can be seen in Figure 9.3, HFRL contains multiple parallel RL agents (we present two agents for briefness) for different coal-fired boiler systems, which may be geographically distributed. The RL agents have the same task of making optimal control of corresponding coal-fired boiler systems. A federated server takes the role of aggregating the models from different RL agents. The basic steps for conducting HFRL can be listed as follows.

- **Step 1:** All participant RL agents train their own RL models according to Figure 9.2 locally and independently, without any exchange of data experience, parameter gradients, and losses.

- **Step 2:** RL agents send their masked model parameters to the server.

- **Step 3:** Federated server encrypts the models from non-identical RL agents and conducts aggregation methods to obtain a federated model.

- **Step 4:** Federated server sends the federated model to the RL agents.

- **Step 5:** RL agents update the local model.

In literature, researchers start to pay attention to studies in HFRL. Liu et al. [2019] proposed Lifelong Federated Reinforcement Learning (LFRL) in the autonomous navigation settings where the main task is to make the robots share their experience so that they can effectively use prior knowledge and quickly adapt to new changes in the environment. The main idea of work can be summarized in three steps.

1. **Independent learning.** Each robot executes its own navigation task in its own environment. Note that the environments can be different, related, or non-related. The basic idea is to conduct lifelong learning locally to learn to avoid diverse types of obstacles.

2. **Knowledge fusion.** The knowledge and skills extracted by the robots from defined or undefined environments are then re-produced by the knowledge fusion process, which produces a final model.

3. **Agent network update.** The parameters of the agents' networks are updated regularly. Thus, the knowledge gained by different agents can be shared through these parameters.

Ren et al. [2019] presented a framework based on the deployment of multiple deep reinforcement learning agents on multiple edge nodes to indicate the decisions of the IoT devices. In order to make better knowledge aggregation when reducing the transmission costs between the IoT devices and edge nodes, the authors employed federated learning to train DRL agents in a distributed fashion. The authors conducted extensive experiments to demonstrate the effectiveness of the proposed scheme of HFRL for distributed IoT devices.

Nadiger et al. [2019] thoroughly described the overall architecture for HFRL, which contains the grouping policy, the learning policy, and the federation policy for participant RL agents. The authors further demonstrated the effectiveness of the proposed architecture based on the Atari game Pong. In the demonstrations, the authors showed that with the proposed approach, there is a median improvement of 17% on the personalization time.

Although the privacy-preserving objective may present more challenges, we can benefit from HFRL in the following ways.

- **To avoid *non-i.i.d.* samples**. It is common that single-task agent may encounter *non-i.i.d.* samples during the learning process. It is clear that one of the main reasons is that for RL tasks with a single-agent setting, the experience gained afterward can be strongly related with previous experience, which may break the *i.i.d.* data assumption. HFRL can provide benefits for building a more accurate and stable reinforcement learning system.

- **To enhance sample efficiency**. Another drawback of conventional RL methods is the poor ability to quickly build stable and accurate models with limited samples (known as low sample efficiency problem), which prevents conventional RL methods from being applied in real-world applications. Under HFRL, we can aggregate knowledge extracted by different agents from non-identical environments to address the low sample efficiency problem.

- **To accelerate the learning process**. Actually, this benefit can be drawn from the above two advantages as a by-product. Combined with the powerful FL framework for aggregating different knowledge learned by non-identical agents, experience from more *non-i.i.d.* samples can accelerate RL learning and achieve better results.

Vertical FRL

Recalling the optimal control problem of coal-fired boiler systems, it is obvious that the working condition of a boiler is not only dependent on the controllable factors, but also on the unobtainable (or unpredictable) factors. For example, the meteorological condition may greatly affect the burning efficiency and steam output of coal-fired boilers. In order to train a more reasonable and robust RL agent, it is natural to extract the knowledge from meteorological data. Unfortunately, professional equipment for real-time and accurate measurements of local meteorological data may be not affordable for small power plants. Moreover, the owner of the power plant may not be interested in the raw meteorological data, but the value extracted from itself. Therefore, in order to train a more robust RL agent, it is natural for the owner of the power plant to cooperate with the meteorological data management department. By return, the meteorological data management department can get paid without directly revealing any real-time meteorological data. This cooperative framework falls into the categorization of VFRL.

In VFRL, there are different RL agents that maintain non-identical observations of the same environment. Each RL agent maintains a corresponding action policy (some agent may have no action policy). The main goal of the cooperative framework is to train a more effective RL agent with the mixed knowledge extracted from the observations of different cooperative agents. During the training or the inference process, any direct transformation of raw data is forbidden. The following presents a possible framework for VFRL—Federated DQN.

As can be seen from Figure 9.4, we name the RL agent who obtains the reward from the environment as Q-network agent (agent A in Figure 9.4), and all other agents as cooperative RL agents:

- **Step 1.** All participant RL agents take actions according to the current environment observations and knowledge extracted. Note that some agents may make no action, which only maintains its own observations of the environment.

- **Step 2.** RL agents obtain the corresponding feedback of the environment, including the current environment observations, the reward, etc.

- **Step 3.** RL agents compute the mid-products by feeding the obtained observations to its neural network and then send the masked mid-products to the Q-network agent.

- **Step 4.** The Q-network agent encrypts all mid-products and trains the Q-network with the current losses through back-propagation.

- **Step 5.** The Q-network agent sends back the masked weight gradients to the cooperative agents.

- **Step 6.** Each cooperative agent encrypts the gradients and updates its own network.

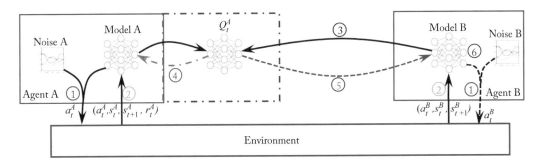

Figure 9.4: Federated-DQN framework.

In literature, existing work falling into VFRL category is Zhuo et al. [2019], which investigates the problem of multi-agent RL system in a cooperative way, when considering the privacy-preserving requirements of agent data, gradients, and models. The FRL framework studied corresponds to the VFRL architecture we presented above (which is named VFRL afterward). The author presented detailed real-life systems where VFRL is meaningful.

Modeling and exploiting the behaviors in systems with multiple cooperative or adversarial agents have long been an interesting challenge for the RL community [Mao et al., 2019, Foerster et al., 2016]. Under the realm of FL, agents may perform heterogeneous tasks, with different states, actions, or rewards (some may have no rewards or actions). The main goal of each VFRL agent is to construct a stable and accurate RL model cooperatively or competitively without direct exchange of experience (including states, actions, and rewards) or the corresponding gradients. Compared with multi-agent RL, the advantages of VFRL can be summarized as follows.

- **To avoid agent and user information-leakage**. In the coal-fired boiler systems, a straight-forward benefit presented for the meteorological data management department is that it can enhance the production efficiency without any leakage of raw real-time meteorological data. This can be cast as a service that can be published to all potential external users.

- **To enhance RL performance**. With proper knowledge extraction methods adopted, we can train a more reasonable and robust RL agent to enhance efficiency. VFRL is advantageous in the sense that it can enable a learning system to leverage such information while preserving privacy.

9.5   CHALLENGES AND OUTLOOK

As an emerging framework for preserving the privacy of different parties and preventing information leakage during the training and inference processes, Federated Learning attracts more and more research attention during the past few years. The following illustrates the challenges together with research directions for FRL.

- *New privacy-preserving paradigms*. Note that the above-cited FRL work adopted either the idea of exchanging parameters or employing Gaussian noise, which is very fragile when faced with adversarial (deceiving) agents or even attackers. More reliable paradigms can be merged into FRL, such as Differential Privacy, Secure Multi-party Computation, Homomorphic Encryption, etc.

- *Transfer FRL*. Although we did not make a single category for Transfer FRL, its importance still urges us to present a meaningful research direction. In conventional RL researches, transferring the experience, knowledge, parameters, or gradients from the already learned tasks to new ones constitutes a research frontier. It is a general common sense in RL community that the goal of learning from prior knowledge is even more challenging than merely learning from samples.

- *FRL Mechanisms*. It can be easily summarized from above-cited work that all existing FRL researches fall into the categorization of Deep Reinforcement Learning. Considering the constraints in the realm of FL, it would be of great meanings in presenting new RL mechanisms (with traditional or DL methods), which constitute another challenging frontier.

CHAPTER 10

Selected Applications

As an innovative modeling mechanism that can build shared and personalized models on decentralized data scattered among multiple parties without compromising user privacy and security, federated learning has promising applications in many important fields such as sales, finance, healthcare, education, urban computing, edge computing, and blockchain, where data cannot be directly aggregated for training machine learning (ML) models due to various reasons. In this chapter, we give an overview of several ongoing and potential applications that could be realized with federated learning beyond the current horizon.

10.1 FINANCE

The financial industry is greatly affected by government regulations in many ways for protecting investors against mismanagement and fraud, sustaining the stability of the financial sector, preserving the privacy and security of user data, and many others. To save cost and workload from government regulations, many financial and banking companies have exploited modern technologies such as AI, cloud services, and mobile Internet technologies to efficiently and effectively provide financial services while complying with strict government regulations.

Take smart consumer financing as an example. The purpose is to leverage ML techniques to offer personalized financial services to creditworthy consumers to encourage consumption. The data features involved in consumer finance mainly include consumer qualification information, purchasing power, and purchasing preference, as well as product characteristics. In practical applications, these data features are likely to be collected by different departments or companies. For example (Figure 10.1), a consumer's qualification information and purchasing power can be inferred from her bank savings, and her purchasing preference for various products or services can be analyzed from her social networks. The characteristics of products are recorded by an e-shop. In this scenario, we are faced with two problems. First, for the protection of user privacy and data security, data barriers between banks, social networking sites, and the e-shopping sites are difficult to break. As a result, the data cannot be directly aggregated. Second, the data stored by the three parties are usually heterogeneous, and traditional ML models cannot directly work on heterogeneous data. For now, these problems have not been efficiently solved with traditional ML methods.

Federated learning and transfer learning are the key to solving these problems. First, based on federated learning, we can build local personalized models for the three parties without exposing their data. Meanwhile, we can leverage transfer learning to address the data heterogene-

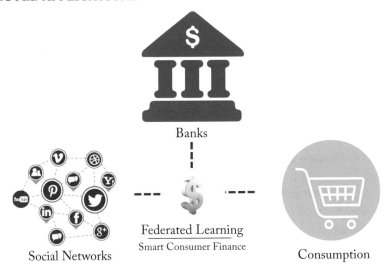

Figure 10.1: Federated learning in smart consumer finance.

ity problem, and overcome the limitations of traditional AI techniques. Therefore, federated learning provides excellent technical support for us to build a cross-enterprise, cross-data and cross-domain ecosystem for big data and AI.

10.2   HEALTHCARE

With the advance in AI technologies, many AI applications have been developed in the medical field with the hope to reduce labor costs and human errors. For example, AI programs for cardiology and radiology have been developed to help diagnose heart diseases and identify cancer cells in the early stages. With the promising applications of health AI, more and more healthcare providers are leveraging AI to create efficiency and improve patient care (Figure 10.2).

However, the adoption of AI technologies in the medical industry remains in its infancy. Existing intelligent medical systems are far from really "intelligence," and some are being questioned for offering unsafe and incorrect treatment recommendations [Chen, 2018, Mearian, 2018]. Many factors may contribute to the deficiency of existing intelligent medical systems. A crucial one is the difficulty in collecting a sufficiently large amount of data with rich features that can comprehensively describe the symptom of a patient. For example, to accurately diagnose a disease, we may need features from various sources including disease symptoms, gene sequences, medical reports, examination results, and academic papers. However, there is no stable data source for filling in values of all those features. Besides, the labels of the majority of the training data are missing. Researchers estimate that it would take 10 years with 10,000 experts to gather a dataset useful enough for developing healthcare AI. The insufficiency of data and

Figure 10.2: Federated learning in intelligent diagnosis.

labels that results in the poor performance of ML models becomes the bottleneck of intelligent medical systems.

To break through this bottleneck, medical institutions could unite together by sharing their data in compliance with privacy protection regulations. Then, we could possess a dataset large enough to train a model that can perform much better than the model trained on data from a single medical institution. Combining federated learning with transfer learning is a promising solution to achieve this goal. First, data from medical institutions are sensitive to privacy and security issues. Directly gathering such data in one location is infeasible. Federated learning allows all participating parties to collaboratively train a shared model without exchanging or exposing their private patient data. Second, transfer learning techniques can help expand the sample and feature space of training data and, in turn, improve the performance of the shared model. Therefore, federated transfer learning can play an important role in the development of intelligent medical systems. If a decent amount of the medical institutions could establish a data alliance together through federated learning in the future, health AI can bring more benefits to more patients.

10.3   EDUCATION

Educators have long called for instructional systems that integrate cross-curricular subjects (e.g., among science, technology, engineering, and mathematics (STEM) subjects and also between STEM and the humanities). However, instructional systems can seldom handle the prerequisite skills, knowledge bases, and experiences necessary to provide such an integrated learning experience. A typical adaptive instructional system (AIS) addresses a single subject at a time, and it often has a unique content knowledge base, adaptive engine, and data management method. For example, a maths AIS knowledge base typically consists of a knowledge graph of granular learning objectives in maths, but it can have many connections to objectives in physics and chemistry. A student's calculus knowledge, for example, could inform their learning experience in physics or chemistry. Thus, an integration of knowledge bases across instructional systems would not only expand the scope of multiple AISs, but also support a richer, cross-curricular adaptive learning experience for students.

To this end, we can encode the knowledge graph of each AIS as a graph neural network proven to have high representational power. Then, we can use federated learning-based approaches to build a comprehensive model that integrates knowledge neural graphs of various AIS, thereby extending the curricular knowledge, the learner model, and the data reach from one AIS to another. In this way, each AIS benefits from data synchronization, latency reduction, and security features of a federated system while maintaining its own knowledge base, adaptive engine, and data.

In addition to integrating educational resources, federated learning can help achieve personalized education (Figure 10.3). More specifically, educational institutions can utilize federated learning to collaboratively build a general learning-plan formulating model based on data stored at students' personal mobile devices such as smartphones, iPads, and laptops. The general model can formulate a standardized learning plan for students having similar backgrounds. On top of that general model, a personalized model that can provide personalized learning instructions can be built for each student based on that student's strengths, needs, skills, and interests.

10.4   URBAN COMPUTING AND SMART CITY

According to Zheng et al. [2013], urban computing is defined as a process of acquiring, integrating, and analyzing big and heterogeneous data generated by a diversity of sources in urban spaces, such as sensors, devices, vehicles, buildings, and humans, for tackling the major issues that cities face, such as air pollution, increased energy consumption, and traffic congestion. It is the technology that helps create smart cities aiming to agilely respond to citizens' needs.

With the development of cloud services, big data, AI, the Internet of Things (IoT) and fifth generation (5G) technologies, smart cities were being built at an increasingly fast pace in many developing and developed countries. After its transient eruption, however, the development of smart cities has entered a slower phase in which cities face many big challenges.

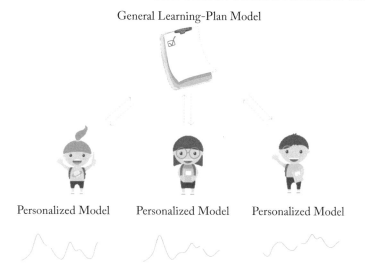

Figure 10.3: Federated learning in education.

In iResearch [2019], four challenges that researchers, engineers, and civil servants have been encountered while building smart cities are summarized.

- Emphasizing on technology while ignoring participation. Emphasizing on informatization and platform construction among large enterprises and institutions, while ignoring the participation of the majority of small businesses.

- Data silos and data fragmentation. The lack of integration of data, applications, and departmental responsibilities for urban management remains unresolved.

- Security risks of intelligent systems. Insufficient attention is being paid to information security, operational security, and network security, which increases city management costs and risks.

- Lack of sustainable model of operation. The market participation mechanism is not comprehensive. Sustainable and just payoff-sharing and reward mechanism regulated by market rules need to be built.

Federated learning with its collaborative and privacy-preserving nature is a promising solution for addressing these challenges (Figure 10.4). Federated learning brings greater opportunities and benefits to small businesses through the collaborative construction of smart city technologies. Under the federated learning, small businesses can collaboratively build intelligent applications by exploiting the data of all participants without compromising privacy and

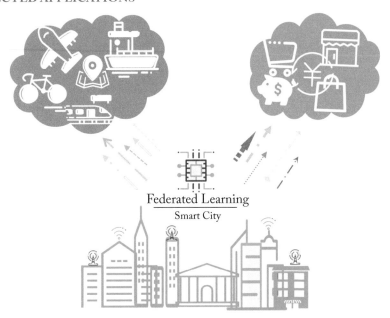

Figure 10.4: Federated learning in urban computing and smart city.

security. For example, by applying federated learning, ride-hailing companies can collaboratively build optimal models to address the vehicle routing problem, not only directly increasing their revenue and improving customer satisfaction, but also gaining side benefits brought by distributing and reducing the traffic congestion.

With federated learning, the issues of data silos can be addressed to some extent. There are many factors that may contribute to data silos, such as regulatory risk, privacy concerns, misaligned incentives, and the high cost of integrating heterogeneous data. In addition to addressing the privacy concerns and bringing benefits to participating parties, federated learning is capable of integrating data with heterogeneous features. For example, current air quality prediction models generally rely on Air Quality Index (AQI) readings from sparsely distributed air quality monitoring stations and meteorological conditions, and are unable to make use of more fine resolution industrial emissions and vehicle exhaust data that have quite different feature spaces from AQI. Vertical federated learning is able to solve this problem, allowing the training of a virtual shared model on data with heterogeneous features.

In order to sustain long-term stability in a data federation and attract more high quality data owners over time, an incentive scheme that shares the profit generated by a federation with participants in a fair and just manner is needed. Federate Learning Incentivizer (FLI) was proposed for this function. We refer interested reader to Chapter 7 for details. The core of this payoff-sharing scheme is to dynamically divides a given budget among data owners in a federation by jointly maximizing the sustainable operation objectives, while minimizing the inequity among the data owners. With FLI , we envisage that more and more data owners will be motivated to contribute high-quality data to the data federation to facilitate the development of smart cities.

10.5   EDGE COMPUTING AND INTERNET OF THINGS

With the soaring number of netizens [CNNIC, 2018, eMarketer, 2017], the popularity of the mobile Internet and mobile phones has promoted the development of Mobile Edge Computing (MEC). MEC allows computing to occur where data are produced (i.e., on the IoT devices) instead of sending data to cloud servers. It can be applied to any single enterprise or institution that has deployed IoT devices, especially mobile devices.

A variety of applications powered by AI techniques (e.g., face recognition, voice assistant, and intelligent background blur) can be deployed on mobile phones. Current solutions for AI applications typically require user data to be uploaded to the cloud server in order to train a giant model. However, this may cause privacy violation and security breach. In addition, with the centralized nature of current AI algorithms, users may suffer from high-latency while using AI apps, especially when connection is weak.

Federated learning allows for building more intelligent models while preserving privacy and security of local data. It can serve as an operating system for edge computing, as it provides the learning protocols for coordination and security (Figure 10.5). More specifically, federated learning enables edge computing devices to collaboratively train an ML model without sending data to the cloud. In addition to the privacy-preserving benefit brought by federated learning, each mobile device ends up with a personalized model that can respond to users immediately. Google has ready tested its federated learning in the Gboard application on Android smart-phones [McMahan and Ramage, 2017]. Their federated learning algorithm utilizes clicking histories of user query suggestions stored on devices to make improvements to the next iteration of Gboard's query suggestion model.

The changes brought about by federated learning are not limited to mobile devices, but smart home terminals as well. Federated learning can make full use of heterogeneous features of data collected by different home devices to build smarter applications. Federated learning models are not only suitable for TV and electric lights, but also can be combined with smart speakers and door lockers for linkage function development, supporting more complex new modes of operation in smart homes.

The bringing model training to edge with federated learning opens up many algorithmic and technical challenges. One of them is that it requires edge devices to be installed with more powerful processors for training complex local models. This requirement will push terminal de-vice makers such as Apple, Huawei, and Xiaomi to develop specialized hardware (e.g., neural processing unit (NPU)) tailored to modern AI techniques such as deep neural networks. With

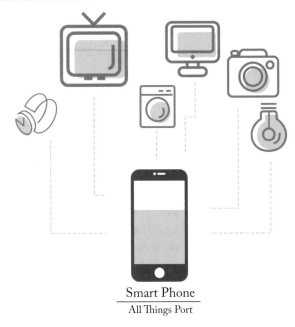

Figure 10.5: Federated learning in edge computing.

the evolution of AI and IoT, AI technologies, and edge computing will not be developing in isolation, but will be moving toward the path of integrated development.

10.6   BLOCKCHAIN

Federated learning provides participants with the capability of collaboratively building powerful ML models and employs privacy-preserving mechanisms to protect the privacy of their data. However, federated learning has been questioned for its vulnerability to backdoor attacks [Bagdasaryan et al., 2019]. For example, malicious participants can poison machines learning models with malicious training samples and undermine the effects of the final model using model replacement techniques. Many secure protocols have been developed to guard against malicious attacks (e.g., defensive distillation and adversarial training regularization). However, in order to actively prevent federated learning from malicious attacks instead of just passive defense, a mechanism that can effectively detect malicious attacks and pinpoint malicious participants is needed.

Blockchain, with its immutability and traceability, can be an effective tool to prevent malicious attacks in federated learning [Preuveneers et al., 2018]. More specifically, the immediate updates made by each participant to its local model can be chained together on the distributed ledger offered by a blockchain such that those model updates are audited. In addition, every model update, be it either local weights or gradients, can be traced to and associated with an individual participant, which helps the detection of tamper attempts and malicious model substitutes. Furthermore, model updates can be chained in a cryptographical way such that their integrity and confidentiality can be guaranteed.

10.7   5G MOBILE NETWORKS

Federated learning has also become an active research topic at the intersection of ML and wireless networks, see, e.g., Habachi et al. [2019], Zhou et al. [2019], Zhu et al. [2018], Samarakoon et al. [2018], Jeong et al. [2018], and particularly for the 5G mobile network and even beyond [Niknam et al., 2019, Letaief et al., 2019, Bennis, 2019, Park et al., 2019]. For instance, the data in wireless networks is usually located at the users and at the network edge, which makes the traditional ML that is based on centralized data collection inapplicable. Federated learning comes as a solution, for addressing not only the data privacy concerns, but also the communication bandwidth, reliability and latency challenges [Bonawitz and Eichner et al., 2019]. Federated learning can also help with building a better wireless network. For example, Niknam et al. [2019] provided an overview of how we can leverage federated learning to address the key challenges and to improve the performance of 5G mobile networks.

CHAPTER 11

Summary and Outlook

In this book, we gave an overview of federated machine learning (a.k.a. federated learning), mainly covering the background information on privacy-preserving machine learning and distributed machine learning, horizontal federated learning, vertical federated learning, federated transfer learning, incentive mechanisms, federated learning in computer vision, natural language processing and recommender systems, federated reinforcement learning, and applications of federated learning in various industrial sectors.

Federated learning was born out of the increasing concern over data fragmentation, data silos, user privacy leakage, and data shortage problems facing machine learning. Our society is becoming aware of the severe impact of user privacy violations by large corporations, and regulators are tightening the laws governing the sharing of private data, such as the most stringent requirements of the GDPR on data security [DLA Piper, 2019]. As the traditional machine learning approaches that are based on centralized data collection is no longer compliant with strict data protection laws, in order for the field of AI to continue advancing, an innovative solution that can preserve data privacy is badly needed.

Federated learning allows multiple parties to hold their own data privately while building a machine learning model collaboratively and securely. With federated learning, data does not need to leave the data owners, and hence privacy can be better protected. In this book, we have discussed several modes in building the federated machine learning model, including horizontal federated learning, vertical federated learning, and federated transfer learning. We have also explored federated reinforcement learning, federated learning in computer vision, natural language processing and federated recommendation systems. As societies move ahead, these techniques are likely to play a major role in moving AI into the next level, a level in which models can be built collaboratively, confidentially and in a privacy-preserving manner. One such example is the Google Gboard system [Bonawitz and Eichner et al., 2019]. In this book, we have also pointed out that federated learning is not only just a technical solution, but also a privacy-preserving ecosystem, such as the FedAI ecosystem [WeBank FedAI, 2019]. Building a federated learning ecosystem is also an economic problem in which incentive mechanisms need be carefully designed to ensure that the profits are shared fairly and transparently. This is so that different parties will want to join the federation in a sustainable way. In addition, such incentive mechanisms should also help federations dissuade malicious participants.

As more application scenarios for federated learning are being explored, the field is becoming ever more inclusive. It spans research and practice in machine learning, statistics, information security, encryption and model compression, game theory and economic principles, incentive mechanism design, and more.

In the future, the federated learning ecosystem is likely to expand. More open-source software will emerge, such as FATE [WeBank FATE, 2019] and PySyft [2019]. Practitioners will be accustomed to building solutions that have all the necessary facets expected by the society, and federated learning will become a prime example of "AI for Social Good."

APPENDIX A

Legal Development on Data Protection

Data and machine learning model sharing has huge benefits in our society, but improper sharing of data can lead to severe privacy infringement. In this appendix, we give three examples of legal developments to address the data privacy issue. We will review some recent regulations from the European Union, USA, and China. This is to provide the readers with more information regarding data protection laws and regulations, and is not intended to be legal advice.

A.1   DATA PROTECTION IN THE EUROPEAN UNION

In the era of big data and AI, concerns about user privacy and data confidentiality are universal. With more and more serious cases of data leakage and privacy breach [Mancuso et al., 2019], data protection is gaining increasing societal attention and public support. The General Data Protection Regulation (GDPR), introduced by the European Union (EU) in 2016 and came into effect in 2018, is currently the most comprehensive and widely adopted data protection laws. GDPR was enacted to protect people residing within the EU from user privacy and data security breaches in the digital age. It is considered as the greatest change to EU user privacy laws in almost 20 years [GDPR Info, 2019, GDPR.ORG, 2019, GDPR overview, 2019].

GDPR replaced the Data Protection Directive (DPD) 95/46/EC [GDPR Info, 2019, GDPR.ORG, 2019, GDPR overview, 2019] in 2016. The EU gave two years for its member states to make sure that GDPR would be fully implementable in each individual member state, and it officially came into effect on May 25, 2018.

GDPR consists of 99 articles that are grouped into 11 chapters, and 173 recitals with explanatory remarks. An outline of GDPR is given below.

- Chapter I presents general provisions, expressed in four articles (Articles 1–4).

- Chapter II outlines the data protection principles, expressed in seven articles (Articles 5–11).

- Chapter III defines the rights of the data subject, expressed in 12 articles (Articles 12–23), which are grouped into five sections.

- Chapter IV defines the rights and obligations of controllers and processors, expressed in 20 articles (Articles 24–43), which are grouped into five sections.

- Chapter V defines the regulations regarding transfers of personal data to third countries or international organizations, expressed in seven articles (Articles 44–50).

- Chapter VI defines the roles of independent supervisory authorities, expressed in nine articles (Articles 51–59), which are grouped into two sections.

- Chapter VII defines the regulations regarding cooperation and consistency, expressed in 17 articles (Articles 60–76), which are grouped into three sections.

- Chapter VIII defines the remedies, liability, and penalties, expressed in eight articles (Articles 77–84).

- Chapter IX defines provisions relating to specific processing situations, expressed in seven articles (Articles 85–91).

- Chapter X defines delegated acts and implementing acts, expressed in two articles (Articles 92 and 93).

- Chapter XI defines delegated acts and implementing acts, expressed in six articles (Articles 94–99).

The official GDPR document and more details can be found at GDPR document [2016]

A.1.1   THE TERMINOLOGY OF GDPR

Article 4 of GDPR clearly defines the terms used. We highlight some of the most important ones here.

- Personal data: Any information relating to an identified or identifiable natural person ("data subject"), such as physical, physiological, genetic, mental, economic, cultural, or social identity of that data subject.

- Processing: Any operation or set of operations which is performed on personal data or on sets of personal data, whether or not by automated means, such as collection, recording, organization, structuring, storage, adaptation or alteration, retrieval, consultation, use, disclosure by transmission, dissemination or otherwise making available, alignment or combination, restriction, erasure, or destruction.

- Cross-border processing: Processing of personal data takes place in more than one member states of EU.

- Profiling: Any form of automated processing of personal data consisting of the use of personal data to evaluate certain personal aspects relating to a natural person.

- Pseudonymisation: The processing of personal data in such a manner that the personal data can no longer be attributed to a specific data subject without the use of additional information, provided that such additional information is kept separately and is subject to technical and organizational measures to ensure that the personal data are not attributed to an identified or identifiable natural person.

- Controller: The natural or legal person, public authority, agency, or other body which, alone or jointly with others, determines the purposes and means of the processing of personal data.

- Processor: A natural or legal person, public authority, agency or other body which processes personal data on behalf of the controller.

- Consent of the data subject: Any freely given, specific, informed, and unambiguous indication of the data subject's wishes by which he or she, by a statement or by a clear affirmative action, signifies agreement to the processing of personal data relating to him or her.

- Personal data breach: A breach of security leading to the accidental or unlawful destruction, loss, alteration, unauthorized disclosure of, or access to, personal data transmitted, stored, or otherwise processed.

A.1.2   HIGHLIGHTS OF GDPR

GDPR enforces strong privacy-preservation rules regarding data processing. We provide here some of the important points.

- **Highlight I: Increased territorial scope (Article 3 of GDPR)**

The increased territorial scope, also known as *extraterritorial applicability*, represents one of the major changes in GDPR as compared to DPD 95/46/EC. Specifically, GDPR applies to the following cases DLA Piper [2019], GDPR document [2016], and TechRepublic [2019].

(i) The processing of personal data by an organization established in the EU, regardless of whether the processing takes place inside the EU or not.

(ii) The processing of personal data of data subjects residing in the EU by an organization that is not established in the EU, where the processing relates to the offering of goods or services to such data subjects in the EU or monitors the behavior of data subjects, as long as their behavior takes place within the EU.

(iii) The processing of personal data of data subjects in the EU by an organization that is not established in the EU, where the processing relates to the monitoring of the behavior of such data subjects in the EU.

(iv) The processing of personal data by an organization that is not established in the EU, but where EU member state law applies by virtue of international public law.

- **Highlight II: Basic principles relating to the processing of personal data (Article 5 of GDPR)**

GDPR provides seven basic principles regarding the processing of personal data [GDPR document, 2016, GDPR Info, 2019, GDPR overview, 2019, Kotsios et al., 2019, University of Groningen, 2019].

(i) Lawfulness, fairness, and transparency: Personal data shall be processed lawfully, fairly and in a transparent manner in relation to the data subject. Transparency implies that any information and communication concerning the processing of personal data must be easily accessible and easy to understand. Clear and plain language needs to be used in this regard. This principle ensures data subjects receive information on the identity of the controllers and the purposes of the processing of personal data.

(ii) Purpose limitation: Personal data shall be collected for specified, explicit, and legitimate purposes, and not further processed in a manner that is incompatible with those purposes.[1]

(iii) Data minimization: Personal data shall be adequate, relevant, and limited to what is necessary in relation to the purposes for which they are processed.

(iv) Accuracy: Personal data shall be accurate and, where necessary, kept up to date. Every reasonable step must be taken to ensure that personal data that are inaccurate, having regard to the purposes for which they are processed, are erased or rectified without any delay.

(v) Storage limitation: Personal data shall be kept in a form which permits identification of data subjects for no longer than is necessary for the purposes for which the personal data are processed.

(vi) Integrity and confidentiality: Personal data shall be processed in a manner that ensures appropriate security of the personal data, including protection against unauthorized or unlawful processing and against accidental loss, destruction or damage, using appropriate technical or organizational measures.

(vii) Accountability: The controller shall be responsible for, and be able to demonstrate compliance with, the six data protection principles (i)–(vi).

---

[1]Further processing for the purposes of the public interest, scientific or historical research or statistical purposes is not considered as incompatible with the initial purposes and is therefore allowed [GDPR overview, 2019, Kotsios et al., 2019, University of Groningen, 2019].

- **Highlight III: Rights of data subjects (Articles 13–22 of GDPR)**

  GDPR defines eight rights for data subjects [GDPR document, 2016, GDPR Info, 2019, GDPR overview, 2019, Kotsios et al., 2019].

  (i) The right to be informed: The right to be informed encompasses your obligation to provide "fair processing information," typically through a privacy notice. It emphasizes the need for transparency over how you use personal data.

  (ii) The right of access: Data subjects have the right to request access to their personal data and to ask how their data are being used by the company after they have been gathered. The company must provide a copy of the personal data, free of charge and in electronic format, if requested.

  (iii) The right to rectification: Data subjects are entitled to have personal data rectified if it is inaccurate or incomplete.

  (iv) The right to erasure (a.k.a. *the right to be forgotten*): Data subjects have the rights to request the deletion or removal of personal data where there is no compelling reason for its continued processing.

  (v) The right to restrict processing: Data subjects can request that their data is not used for processing. Their data record can remain in place, but not be used.

  (vi) The right to data portability: Individuals have the right to transfer their data from one service provider to another. It must happen in a commonly used and machine readable format.

  (vii) The right to object: Data subjects have the rights to stop the processing of their data for direct marketing. There are no exemptions to this rule, and any processing must stop as soon as the request is received. In addition, this right must be made clear to data subjects at the very beginning of any communication.

  (viii) Rights in relation to automated decision making and profiling: Data subjects have the rights not to be subject to decision-making when it is based on automated processing and it produces a legal effect or a similarly significant effect on the data subjects.

- **Highlight IV: Data protection by design and by default (Article 25 of GDPR)**

  Under GDPR, the controller has a general obligation to implement technical and organizational measures (e.g., pseudonymisation and data minimization) to show that it has considered and integrated data protection into the processing activities. The controller shall implement appropriate technical and organizational measures for ensuring that, by default, only personal data which are necessary for each specific purpose of the processing are processed.

- **Highlight V: Breach notification (Article 33 of GDPR)**

GDPR demands that all organizations report certain types of data breach to the relevant supervisory authority, and in some cases to the individuals affected. Under GDPR, breach notifications are now mandatory in all member states where a data breach is likely to "result in a risk for the rights and freedoms of individuals." This must be done within 72 hours of first becoming aware of the breach. Data processors are also required to notify their customers, the controllers, "without undue delay" after first becoming aware of a data breach.

- **Highlight VI: Administrative fines in respect of infringements of GDPR (Article 83 of GDPR)**

Under GDPR, fines for breaches of certain important provisions can amount to up to €20 million or 4% of global annual turnover, whichever is the higher. Fines for breaches of other provisions can amount to up to €10 million or 2% of global annual turnover, whichever is higher. The fines under GDPR are significantly higher than those which can be imposed under current laws (e.g., up to £550,000 under current British laws).

The maximum fine is imposed for the most serious infringements (e.g., not having sufficient customer consent to process data or violating the core of "Privacy by Design" concepts). There is a tiered approach to fines. For example, a company can be fined 2% of its global annual turnover for not having their records in order (cf. Article 28 of GDPR), not notifying the supervising authority and data subject about a breach, or not conducting impact assessment. It is important to note that these rules apply to both controllers and processors, meaning "clouds" are not exempt from GDPR enforcement.

A.1.3   IMPACT OF GDPR

GDPR gives customers, contractors, and employees more power over their data, and less power to the organizations that collect and use such data. Under GDPR, organizations must ensure that data subjects are able to obtain human intervention of autonomous decision-making, as well as to obtain an explanation of the automated decision-making and challenge it. The impact of GDPR is far-reaching. Overall, GDPR is very much in favor of individual data owners. The new regulations that have been implemented allow users to discover who have their data, why they have them, where they are stored and who are accessing them [McGavisk, 2019].

The positive implications of GDPR include [McGavisk, 2019] the following.

- Improved cybersecurity: While GDPR has direct impact on user privacy and data security, it also encourages organizations to develop and improve cybersecurity measures, mitigating the risks of any potential data leakage.

- Standardization of data protection: GDPR ensures that once an organization is GDPR compliant, it is free to operate throughout EU without being required to deal with individual data protection legislation of each member state.

• Brand safety: If an organization can become a trusted holder of information in line with GDPR, it stands a better chance to create a long-lasting and loyal relationship with customers.

The possible negative implications of GDPR are [McGavisk, 2019] as follows.

• Non-compliance penalties: The consequence of non-compliance is overwhelming and it has encouraged organizations to make more efforts to consider their data protection responsibilities inside the EU.

• The cost of compliance: most organizations started by instating a Data Protection Officer to take responsibility for ensuring internal policies were updated and any required processes were implemented.

• Over-regulation: Adding a double opt-in inside a form presents the modern customer with a never-ending message of consent. The continuous presence of opting-in may discourage some customers from registering as they delay the requirement of opting-in until they are absolutely certain of their interest.

The impact of GDPR on the AI industry is profound. For building machine learning models, we face the challenge that our data are stored in isolated silos, but we may be forbidden in many situations to collect and transfer data for processing [Yang et al., 2019]. That is to say, GDPR makes data collection harder, if not impossible. For AI applications with respect to data processing that has direct legal effects on customers, such as credit applications and workplace monitoring, GDPR will limit the usefulness of AI for these purposes. For example, under Article 22 and Recital 71, a business would generally need to undergo the time-consuming process of obtaining and recording explicit consent from all customers involved [Roe, 2018]. Note that, even with federated learning, we need to obtain explicit consent from the users before carrying out federated model training and we need to explain clearly what the data is used for, in order to be compliant with the GDPR.

A.2   DATA PROTECTION IN THE USA

Unlike the EU, there is no single law or regulation for general data protection in the United States of America (USA). In the USA, there are several sector-specific and medium-specific national user privacy and data security laws, including laws and regulations that apply to financial institutions, telecommunications companies, personal health information, credit report information, children's information, telemarketing, and direct marketing [DLA Piper, 2019, Pierce, 2018]. The USA also has hundreds of user privacy and data security laws among its 50 states and territories, such as requirements for safeguarding data, disposal of data, privacy policies, appropriate use of social security numbers and data breach notification.

Privacy law in the USA is a complex patchwork of national privacy laws and regulations that address particular issues or sectors, and state laws that address privacy and security of personal information, as well as federal and state prohibitions against unfair or deceptive data usage [DLA Piper, 2019].

One representative example is the state of California. California alone has more than 25 state user privacy and data security laws, including the California Consumer Privacy Act (CCPA) enacted in 2018, which will take effect on January 1, 2020 [CCPA, 2019]. CCPA applies across sectors, introduces sweeping definitions and broad individual rights, and imposes substantial requirements and restrictions on the collection, use and disclosure of personal information. CCPA grants consumers the right to know what information is collected and who it is shared with. Consumers will have the option of barring tech companies from selling their data [CCPA, 2019].

The Federal Trade Commission (FTC) of the USA has jurisdiction over most commercial entities and has authority to issue and enforce privacy regulations in specific areas, such as telemarketing, commercial email, and children's privacy. The FTC can take enforcement actions to protect consumers against unfair or deceptive trade practices, including materially unfair privacy and data security practices. Further, a wide range of sector-specific regulators, such as in health care, financial services, telecommunications, and insurance, also have authority to issue and enforce user privacy and data security regulations, with respect to the entities under their jurisdiction [DLA Piper, 2019].

A.3   DATA PROTECTION IN CHINA

There has been a boom of AI research and commercialization in China in the past few years, which is partly due to the strong support from the central Chinese government. While making great efforts in promoting AI, the Chinese government has also introduced new laws and regulations for data protection. The Cyberspace Administration of China (CAC) is currently considered as the primary data protection authority in the People's Republic of China (PRC), and there are also enforcement regulators such as the Ministry of Public Security, and sector-specific regulators that may monitor and enforce data protection issues, such as the People's Bank of China and China Banking Regulatory Commission that regulate banks and financial institutions [DLA Piper, 2019].

Similar to the USA, there is no single comprehensive data protection law in China. The General Principles of Civil Law of the PRC has generally interpret data protection rights as a right of reputation or right of privacy [DLA Piper, 2019]. Rules and regulations relating to data protection and data security are part of a complex framework and are found across various laws and regulations [DLA Piper, 2019]. The following are a few examples.

The PRC Consumer Rights Protection Law (also known as Consumer Protection Law), which became effective from March 15, 2014, contains data protection obligations that are applicable to most if not all types of businesses that deal with consumers' personal data. The Consumer Protection Law was further supplemented by the Measures on Penalties for Infringing Upon the Rights and Interests of Consumers, effective from March 15, 2015. In addition, the draft Implementation Regulations for the PRC Consumer Protection Law released on August 5, 2016, reiterate and clarify some of the data protection obligations with regards to consumers' personal information [DLA Piper, 2019].

The PRC Cyber Security Law, enacted in June 1, 2017, is the first national-level law to address cyber security and data protection. It requires that Internet businesses must not leak or tamper with personal information that they collect and that, when conducting data transactions with third parties, they need to ensure that the proposed contract follow legal data protection obligations [DLA Piper, 2019, Yang et al., 2019].

To implement the PRC Cyber Security Law, on January 2, 2018, China issued the national standard for the protection of personal information (GB/T 35273-2017 Information Technology—Personal Information Security Specification), also known as PIS Specification in short [Mancuso et al., 2019, GB688, 2019], which entered into force on May 1, 2018. This standard (although not legally binding) sets out the best practices that will be expected by regulators who audit companies and enforce China's existing data protection rules [Mancuso et al., 2019, Pierce, 2018].

The E-Commerce Law of the PRC, which was passed in August 2018 and became effective from January 1, 2019, enforces the requirements to protect personal information in an e-commerce context. This new law, which aims to help clean up China's reputation as a major source of counterfeit and knock-off merchandise, also addresses other important aspects of e-commerce, including false advertising, consumer protection, data protection and cyber security. The framework of this new law is comprehensive. For example, there are chapters and articles covering data protection and promotion of consumer protection, and the provision of substantial civil and criminal penalties regarding data security breaches. The E-Commerce Law will make it more difficult for e-commerce companies to develop added-value from their customers' personal data that are collected.

The National Health Commission of the PRC issued the Administrative Measures on the Standards, Security and Service of National Health and Medical Big Data on September 14, 2018 (shorted as the Measures). Under the Measures, health and medical big data refers to health and medical related data that is generated from the process of disease control and prevention or health management. Medical institutions and related entities are responsible organizations for the security and application management of health and medical data. It is required that health and medical data should be securely stored on reliable servers within the territory of China. If health and medical data needs to be transferred overseas, the responsible organizations must undertake a security evaluation procedure when selecting a service agent. The Responsible organizations shall ensure that the agent complies with the relevant requirements and jointly undertake responsibility with the selected agent. Further, China is also in the process of establishing rules for cross-border transfer of personal information and important data via draft Measures for Security Assessment of Cross-border Transfer of Personal Information and Important Data, and draft Guidelines for Data Cross-Border Transfer Security Assessment [Shah et al., 2019].

Finally, as AI is developing fast in China, new data protection laws and regulations are also continuously emerging. For instance, on May 28, 2019, the CAC released the Draft Measures for Data Security Management (shorted as Draft Measures) for public comment. The comment period ended on June 28, 2019. The release of these Draft Measures demonstrates China's continuing efforts to implement the data protection requirements imposed by China's Cyber Security Law [Covington and Burling, 2019]. This new Draft Measures incorporate some of personal information protection requirements specified in the Standard and the Draft Amendment (i.e., GB/T 35273-2017), and also introduce a number of new requirements for the protection of "important data," which is defined as "data that, if leaked, may directly affect China's national security, economic security, social stability, or public health and security." These Draft Measures serve as reinforcement of the PRC Cyber Security Law.

Bibliography

J. Ren, H. Wang, T. Hou, et al., Federated learning-based computation offloading optimization in edge computing-supported Internet of things, *IEEE Access*, 7:69194–69201, June 2019. `https://ieeexplore.ieee.org/stamp/stamp.jsp?arnumber=8728285` DOI: 10.1109/access.2019.2919736. 128

C. Nadiger, A. Kumar, and S. Abdelhak, Federated reinforcement learning for fast personalization, In *Proc. of IEEE 2nd International Conference on Artificial Intelligence and Knowledge Engineering (AIKE)*, August 2019. DOI: 10.1109/aike.2019.00031. 128

Carnegie Mellon University, LEAF: A benchmark for federated settings, July 2019. `https://leaf.cmu.edu/` and `https://github.com/TalwalkarLab/leaf` 13

S. Caldas, P. Wu, T. Li, et al., LEAF: A benchmark for federated settings, January 2019. `https://arxiv.org/abs/1812.01097` 13

T. Li, A. K. Sahu, M. Zaheer, et al., Federated optimization for heterogeneous networks, July 2019. `https://arxiv.org/abs/1812.06127` 10, 33, 55, 56

D. Song, Decentralized federated learning, June 2019. `https://drive.google.com/file/d/1Bk3ldYJcYo405uwATsqC8ZD1_UcLGlRL/view` 65

N. Hynes, R. Cheng, and D. Song, Efficient deep learning on multi-source private data, July 2018. `https://arxiv.org/abs/1807.06689` 65

N. Agarwal, A. T. Suresh, F. Yu, et al., cpSGD: Communication-efficient and differentially-private distributed SGD, May 2018. `https://arxiv.org/abs/1805.10559` 64

K. Pillutla, S. M. Kakade, and Z. Harchaoui, Robust aggregation for federated learning, May 2019. `https://krishnap25.github.io/papers/2019_rfa.pdf` 64

L. Melis, C. Song, E. D. Cristofaro, et al., Exploiting unintended feature leakage in collaborative learning, November 2018. `https://arxiv.org/abs/1805.04049` DOI: 10.1109/sp.2019.00029. 10, 64

M. Mohri, G. Sivek, and A. T. Suresh, Agnostic federated learning, February 2019. `https://arxiv.org/abs/1902.00146` 64

Y. Ma, X. Zhu, and J. Hsu, Data poisoning against differentially-private learners: Attacks and defenses, March 2019. `https://arxiv.org/abs/1903.09860` DOI: 10.24963/ijcai.2019/657. 64

T. D. Nguyen, S. Marchal, M. Miettinen, et al., DÏoT: A federated self-learning anomaly detection system for IoT, May 2019. `https://arxiv.org/abs/1804.07474` 65

C. Xie, S. Koyejo, and I. Gupta, Asynchronous federated optimization, May 2019. `https://arxiv.org/abs/1903.03934` 10, 65

J. Wang and G. Joshi, Adaptive communication strategies to achieve the best error-runtime trade-off in local-update SGD, March 2019. `https://arxiv.org/abs/1810.08313` 64

V. Zantedeschi, A. Bellet, and M. Tommasi, Fully decentralized joint learning of personalized models and collaboration graphs, June 2019. `https://arxiv.org/abs/1901.08460` 53, 65

Google Workshop on Federated Learning and Analytics, June 2019. `https://sites.google.com/view/federated-learning-2019/home` 64, 65

N. Srivastava, G. Hinton, A. Krizhevsky, et al., Dropout: A simple way to prevent neural networks from overfitting, *Journal of Machine Learning Research*, 15:1929–1958, June 2014. 60

O. Gupta and R. Raskar, Distributed learning of deep neural network over multiple agents, *Journal of Network and Computer Applications*, 116:1–8, August 2018. `https://arxiv.org/abs/1810.06060v1` DOI: 10.1016/j.jnca.2018.05.003. 7

A. W. Trask, *Grokking Deep Learning*, Manning Publications, February 2019. 1, 33

P. Vepakomma, O. Gupta, T. Swedish, et al., Split learning for health: Distributed deep learning without sharing raw patient data, In *ICLR Workshop on AI for Social Good*, May 2019. `https://splitlearning.github.io/` 7, 8

T.-Y. Liu, W. Chen, and T. Wang, Distributed machine learning: Foundations, trends, and practices, In *Proc. of the 26th International Conference on World Wide Web Companion (WWW Companion)*, April 2017. DOI: 10.1145/3041021.3051099. 33

E. Hesamifard, H. Takabi, and M. Ghasemi, CryptoDL: Deep neural networks over encrypted data, *ArXiv Preprint ArXiv:1711.05189*, November 2017. `https://arxiv.org/abs/1711.05189` 29, 90

R. Thibaux and M. I. Jordan, Hierarchical beta processes and the Indian buffet process, In *Proc. of 11th International Workshop on Artificial Intelligence and Statistics*, pp. 564–571, 2007. 110

M. Yurochkin, M. Agarwal, S. Ghosh, et al., Bayesian nonparametric federated learning of neural networks, *ArXiv Preprint ArXiv:1905.12022*, May 2019. `https://arxiv.org/abs/1905.12022` 110

A. G. Roy, S. Siddiqui, S. Pölsterl, et al., Braintorrent: A peer-to-peer environment for decentralized federated learning, *ArXiv Preprint ArXiv:1905.06731*, May 2019. https://arxiv.org/abs/1905.06731 110

M. J. Sheller, G. A. Reina, B. Edwards, et al., Multi-institutional deep learning modeling without sharing patient data: A feasibility study on brain tumor segmentation, In *International MICCAI Brainlesion Workshop*, pp. 92–104, Springer, 2018. DOI: 10.1007/978-3-030-11723-8_9. 11, 109

S. Silva, B. Gutman, E. Romero, et al., Federated learning in distributed medical databases: Meta-analysis of large-scale subcortical brain data, *ArXiv Preprint ArXiv:1801.08553*, October 2018. https://arxiv.org/abs/1810.08553 DOI: 10.1109/isbi.2019.8759317. 110

I. Augenstein, S. Ruder, and A. Søgaard, Multi-task learning of pairwise sequence classification tasks over disparate label spaces, *ArXiv Preprint ArXiv:1802.09913*, February 2018. https://arxiv.org/abs/1802.09913 DOI: 10.18653/v1/n18-1172. 114

X. Chen and C. Cardie, Multinomial adversarial networks for multi-domain text classification, *ArXiv Preprint ArXiv:1802.05694*, February 2018. https://arxiv.org/abs/1802.05694 DOI: 10.18653/v1/n18-1111. 114

K. Cho, B. van Merriënboer, C. Gulcehre, et al., Learning phrase representations using RNN encoder-decoder for statistical machine translation, *ArXiv Preprint ArXiv:1406.1078*, June 2014. https://arxiv.org/abs/1406.1078 DOI: 10.3115/v1/d14-1179. 111

S. Hochreiter and J. Schmidhuber, Long short-term memory, *Neural Computation*, 9(8):1735–1780, November 1997. DOI: 10.1162/neco.1997.9.8.1735. 111

S. Ji, S. Pan, G. Long, et al., Learning private neural language modeling with attentive aggregation, *ArXiv Preprint ArXiv:1812.07108*, December 2018. https://arxiv.org/abs/1812.07108 DOI: 10.1109/ijcnn.2019.8852464. 113

D. Leroy, A. Coucke, T. Lavril, et al., Federated learning for keyword spotting, In *IEEE International Conference on Acoustics, Speech and Signal Processing (ICASSP)*, pp. 6341–6345, 2019. DOI: 10.1109/icassp.2019.8683546. 113

H. B. McMahan, D. Ramage, K. Talwar, et al., Learning differentially private recurrent language models, *ArXiv Preprint ArXiv:1710.06963*, October 2017. https://arxiv.org/abs/1710.06963 32

S. Ruder and B. Plank, Strong baselines for neural semi-supervised learning under domain shift, *ArXiv Preprint ArXiv:1804.09530*, April 2018. https://arxiv.org/abs/1804.09530 DOI: 10.18653/v1/p18-1096.

S. Ruder, I. Vulić, and A. Søgaard, A survey of cross-lingual word embedding models, *ArXiv Preprint ArXiv:1706.04902*, June 2017. https://arxiv.org/abs/1706.04902 DOI: 10.1613/jair.1.11640. 114

S. Zhang, L. Yao, A. Sun, et al., Deep learning based recommender system: A survey and new perspectives, *ACM Computing Surveys*, 52(1):5:1–5:38, 2019. DOI: 10.1145/3285029. 115

G. Adomavicius and A. Tuzhilin, Toward the next generation of recommender systems: A survey of the state-of-the-art and possible extensions, *IEEE Transactions on Knowledge and Data Engineering*, 17(6):734–749, 2005. DOI: 10.1109/tkde.2005.99. 115

Y. Zhou, D. M. Wilkinson, R. Schreiber, et al., Large-scale parallel collaborative filtering for the netflix prize, In *Proc. of 4th International Conference Algorithmic Aspects in Information and Management (AAIM)*, pp. 337–348, June 2018. DOI: 10.1007/978-3-540-68880-8_32. 116

E. Kharitonov, Federated online learning to rank with eution strategies, In *Proc. of the 12th ACM International Conference on Web Search and Data Mining*, (9):249–257, February 2019. DOI: 10.1145/3289600.3290968. 118

M. Ammad-ud-din, E. Ivannikova, S. A. Khan, et al., Federated collaborative filtering for privacy-preserving personalized recommendation system, *ArXiv Preprint ArXiv:1901.09888*, January 2019. http://arxiv.org/abs/1901.09888 11, 116

J. Trienes, A. T. Cano and D. Hiemstra, Recommending users: Whom to follow on federated social networks, *ArXiv Preprint ArXiv:1811.09292*, November 2018. http://arxiv.org/abs/1811.09292 118

H. Mao, Z. Zhang, Z. Xiao, et al., Modelling the dynamic joint policy of teammates with attention multi-agent DDPG, In *Proc. of the 18th International Conference on Autonomous Agents and MultiAgent Systems*, pp. 1108–1116, July 2019. 130

J. Foerster, I. A. Assael, N. de Freitas, et al., Learning to communicate with deep multi-agent reinforcement learning, In *Advances in Neural Information Processing Systems*, pp. 2137–2145, 2016. 130

G. Barth-Maron, M. W. Hoffman, D. Budden, et al., Distributed distributional deterministic policy gradients, *ArXiv Preprint ArXiv:1804.08617*, April 2018. http://arxiv.org/abs/1804.08617

L. Espeholt, H. Soyer, R. Munos, et al., Impala: Scalable distributed deep-RL with importance weighted actor-learner architectures, *ArXiv Preprint ArXiv:1802.01561*, February 2018. http://arxiv.org/abs/1802.01561

R. M. Kretchmar, Parallel reinforcement learning, In *The 6th World Conference on Systemics, Cybernetics, and Informatics*, 2002. 127

M. Grounds and D. Kudenko, Parallel reinforcement learning with linear function approxima-
tion, In *Proc. of the 5th, 6th, and 7th European Conference on Adaptive and Learning Agents and
Multi-agent Systems: Adaptation and Multi-agent Learning*, pp. 60–74, Springer-Verlag, 2008.
DOI: 10.1007/978-3-540-77949-0_5. 127

V. Mnih, K. Kavukcuoglu, D. Silver, et al., Human-level control through deep reinforcement
learning, *Nature*, 518:529–533, February 2015. DOI: 10.1038/nature14236.

B. Liu, L. Wang, M. Liu, et al., Lifelong federated reinforcement learning: A learning architecture for navigation in cloud robotic systems, *ArXiv Preprint ArXiv:1901.06455*, January
2019. http://arxiv.org/abs/1901.06455 DOI: 10.1109/lra.2019.2931179. 128

H. Zhu and Y. Jin, Multi-objective eutionary federated learning, *ArXiv Preprint
ArXiv:1812.07478v2*, December 2018. http://arxiv.org/abs/1812.07478v2 47

R. Cramer, I. Damgård, and J. B. Nielsen, Multiparty computation from threshold homomorphic encryption, B. Pfitzmann, Ed., *EUROCRYPT 2001 (LNCS)*, 2045:280—299, Springer,
Heidelberg, 2001. DOI: 10.1007/3-540-44987-6_18. 22

M. Keller, E. Orsini, and P. Scholl, Mascot: Faster malicious arithmetic secure computation with
oblivious transfer, In *Proc. of the ACM SIGSAC Conference on Computer and Communications
Security (CSS)*, pp. 830–842, October 2016. DOI: 10.1145/2976749.2978357. 22, 23, 25, 48

M. Keller, V. Pastro, and D. Rotaru, Overdrive: Making SPDZ great again, In J.B. Nielsen
and V. Rijmen, Eds., *Advances in Cryptology—EUROCRYPT*, pp. 158–189, Cham, Springer
International Publishing, 2018. DOI: 10.1007/978-3-319-78372-7. 25, 48

I. Damgård, D. Escudero, T. Frederiksen, et al., New primitives for actively-secure MPC over
rings with applications to private machine learning, *IACR Cryptology ePrint Archive*, 2019.
DOI: 10.1109/sp.2019.00078. 26

M. Abadi, A. Chu, I. Goodfellow, et al., Deep learning with differential privacy, In *Proc. of
the ACM SIGSAC Conference on Computer and Communications Security*, pp. 308–318, 2016.
DOI: 10.1145/2976749.2978318. 32, 43, 51

Y. Aono, T. Hayashi, L. Trieu Phong, et al., Scalable and secure logistic regression via homomorphic encryption, In *Proc. of the 6th ACM Conference on Data and Application Security and
Privacy*, pp. 142–144, 2016. DOI: 10.1145/2857705.2857731. 43, 88

H. Bae, J. Jang, D. Jung, et al., Security and privacy issues in deep learning, December 2018.
https://arxiv.org/abs/1807.11655

K. Bonawitz, V. Ivanov, B. Kreuter, et al., Practical secure aggregation for federated learning on
user-held data, *ArXiv Preprint ArXiv:1611.04482*, November 2016. http://arxiv.org/ab
s/1611.04482 44, 48, 89

K. Chaudhuri and C. Monteleoni, Privacy-preserving logistic regression, In *Advances in neural information processing systems*, pp. 289–296, 2009. 42

W. Du and Z. Zhan, Building decision tree classifier on private data, In *Proc. of the IEEE International Conference on Privacy, Security and Data Mining-Ume 14*, pp. 1–8, Australian Computer Society, Inc., 2002. 41

C. Dwork, Differential privacy: A survey of results, In *Theory and Applications of Models of Computation, 5th International Conference, TAMC, Proceedings*, pp. 1–19, Xi'an, China, April 2008. DOI: 10.1007/978-3-540-79228-4_1. 42

C. Dwork, A firm foundation for private data analysis, *Communications of the ACM*, 54(1):86–95, 2011. DOI: 10.1145/1866739.1866758. 43

C. Dwork and A. Roth, The algorithmic foundations of differential privacy, *Foundations and Trends® in Theoretical Computer Science*, 9(3–4):211–407, 2014. DOI: 10.1561/0400000042. 31

W. Fang and B. Yang, Privacy preserving decision tree learning over vertically partitioned data, In *IEEE International Conference on Computer Science and Software Engineering*, 3:1049–1052, 2008. DOI: 10.1109/csse.2008.731. 10, 41

S. E. Fienberg, W. J. Fulp, A. B. Slavkovic, et al., Secure log-linear and logistic regression analysis of distributed databases, In *Proc. of International Conference on Privacy in Statistical Databases*, pp. 277–290, Springer, 2006. DOI: 10.1007/11930242_24. 43

G. Jagannathan, K. Pillaipakkamnatt, and R. N. Wright, A practical differentially private random decision tree classifier, In *Proc. of IEEE International Conference on Data Mining Workshops*, pp. 114–121, 2009. DOI: 10.1109/icdmw.2009.93. 42

X. Lin, C. Clifton, and M. Zhu, Privacy-preserving clustering with distributed EM mixture modeling, *Knowledge and Information Systems*, 8(1):68–81, 2005. DOI: 10.1007/s10115-004-0148-7. 44

Y. Lindell and B. Pinkas, Privacy preserving data mining, *Journal of Cryptology*, 15(3), 2002. DOI: 10.1007/s00145-001-0019-2. 25, 41

M. Liu, H. Jiang, J. Chen, et al., A collaborative privacy-preserving deep learning system in distributed mobile environment, In *International Conference on Computational Science and Computational Intelligence (CSCI)*, pp. 192–197, 2016. DOI: 10.1109/csci.2016.0043. 10, 44

O. L. Mangasarian, E. W. Wild, and G. M. Fung, Privacy-preserving classification of vertically partitioned data via random kernels, *ACM Transactions on Knowledge Discovery from Data (TKDD)*, 2(3):12, 2008. DOI: 10.1145/1409620.1409622. 10, 43

R. Mendes and J. P. Vilela, Privacy-preserving data mining: Methods, metrics, and applications, *IEEE Access*, 5:10562–10582, 2017. DOI: 10.1109/access.2017.2706947. 10, 17, 20, 44

P. Mohassel and Y. Zhang, SecureML: A system for scalable privacy-preserving machine learning, In *Proc. of Symposium on Security and Privacy (SP)*, pp. 19–38, 2017. DOI: 10.1109/sp.2017.12. 10, 23, 25, 44, 48, 74

N. Papernot, M. Abadi, U. Erlingsson, et al., Semi-supervised knowledge transfer for deep learning from private training data, *ArXiv Preprint ArXiv:1610.05755*, October 2016. http://arxiv.org/abs/1610.05755 31

N. Papernot, S. Song, I. Mironov, et al., Scalable private learning with pate, *ArXiv Preprint ArXiv:1802.08908*, February 2018. http://arxiv.org/abs/1802.08908 31

M. Park, J. Foulds, K. Chaudhuri, et al., DP-EM: Differentially private expectation maximization, *ArXiv Preprint ArXiv:1605.06995*, May 2016. http://arxiv.org/abs/1605.06995 43

J. Quinlan, Induction of decision trees, *Machine Learning*, pp. 81–106, March 1986. DOI: 10.1007/bf00116251. 41

R. L. Rivest, L. Adleman, and M. L. Dertouzos, On data banks and privacy homomorphisms, *Foundations of Secure Computation*, 4(11):169–180, 1978. 19, 26

A. Shamir, How to share a secret, *Communications of the ACM*, 22(11):612–613, 1979. DOI: 10.1145/359168.359176. 22, 24

A. B. Slavkovic, Y. Nardi, and M. M. Tibbits, Secure logistic regression of horizontally and vertically partitioned distributed databases, In *7th International Conference on Data Mining Workshops (ICDMW)*, pp. 723–728, 2007. DOI: 10.1109/icdmw.2007.114. 44

S. Song, K. Chaudhuri, and A. D. Sarwate, Stochastic gradient descent with differentially private updates, In *Global Conference on Signal and Information Processing*, pp. 245–248, 2013. DOI: 10.1109/globalsip.2013.6736861. 43

J. Vaidya and C. Clifton, Privacy preserving naive Bayes classifier for vertically partitioned data, In *Proc. of the SIAM International Conference on Data Mining*, pp. 522–526, 2004. DOI: 10.1137/1.9781611972740.59. 10, 44

P. Vepakomma, T. Swedish, R. Raskar, et al., No peek: A survey of private distributed deep learning, *ArXiv Preprint ArXiv:1812.03288*, December 2018. vailable: http://arxiv.org/abs/1812.03288 7, 10, 40, 44

K. Wang, Y. Xu, R. She, et al., Classification spanning private databases, In *Proc. of the National Conference on Artificial Intelligence*, 21:293, Menlo Park, CA, Cambridge, MA, London, AAAI Press, MIT Press, 1999, 2006. 41

E. Wild and O. Mangasarian, Privacy-preserving classification of horizontally partitioned data via random kernels, *Technical Report*, 2007. 10, 43

K. Xu, H. Yue, L. Guo, et al., Privacy-preserving machine learning algorithms for big data systems, In *Proc. of 35th International Conference on Distributed Computing Systems*, pp. 318–327, 2015. DOI: 10.1109/icdcs.2015.40. 10, 44

S. Yakoubov, A gentle introduction to Yao's garbled circuits, 2017. http://web.mit.edu/sonka89/www/papers/2017ygc.pdf 23

A. C. Yao, Protocols for secure computations, In *FOCS*, 82:160–164, 1982. DOI: 10.1109/sfcs.1982.38. 19, 28

A. C.-C. Yao, How to generate and exchange secrets, In *Proc. of 27th Annual Symposium on Foundations of Computer Science*, pp. 162–167, 1986. DOI: 10.1109/sfcs.1986.25. 21, 23

H. Yu, X. Jiang, and J. Vaidya, Privacy-preserving SVM using nonlinear kernels on horizontally partitioned data, In *Proc. of the ACM Symposium on Applied Computing*, pp. 603–610, 2006. DOI: 10.1145/1141277.1141415. 44

J. Zhan and S. Matwin, Privacy-preserving support vector machine classification, *International Journal of Intelligent Information and Database Systems*, 1(3–4):356–385, 2007. DOI: 10.1504/ijiids.2007.016686. 44

D. Zhang, X. Chen, D. Wang, et al., A survey on collaborative deep learning and privacy-preserving, In *3rd International Conference on Data Science in Cyberspace (DSC)*, pp. 652–658, 2018. DOI: 10.1109/dsc.2018.00104. 10, 43, 44

L. Song, J. Mao, Y. Zhuo, et al., HyPar: Towards hybrid parallelism for deep learning accelerator array, In *Proc. of 25th International Symposium on High-Performance Computer Architecture*, February 2019. https://arxiv.org/abs/1901.02067 DOI: 10.1109/hpca.2019.00027. 40

A. Krizhevsky, One weird trick for parallelizing conutional neural networks, *ArXiv Preprint ArXiv:1404.5997*, April 2014. https://arxiv.org/abs/1404.5997 40

M. Wang, C.-C. Huang, and J. Li, Unifying data, model and hybrid parallelism in deep learning via tensor tiling, *ArXiv Preprint ArXiv:1805.04170*, May 2018. https://arxiv.org/abs/1805.04170 40

N. Pansare, M. Dusenberry, N. Jindal, et al., Deep learning with Apache SystemML, *ArXiv Preprint ArXiv:1802.04647*, February 2018. https://arxiv.org/abs/1802.04647 39

D. Shrivastava, S. Chaudhury, and Dr. Jayadeva, A data and model-parallel, distributed and scalable framework for training of deep networks in Apache Spark, *ArXiv Preprint ArXiv:1708.05840*, August 2017. https://arxiv.org/abs/1708.05840 40

M. Boehm, S. Tatikonda, B. Reinwald, et al., Hybrid parallelization strategies for large-scale machine learning in SystemML, In *Proc. of the VLDB Endowment*, pp. 553–564, March 2016. DOI: 10.14778/2732286.2732292. 39

Apache Hadoop MapReduce, June 2019. `https://hadoop.apache.org/docs/r2.8.0/hado op-mapreduce-client/hadoop-mapreduce-client-core/MapReduceTutorial.html` 37, 39

Apache Hadoop YARN, June 2019. `https://hadoop.apache.org/docs/current/hadoop-yarn/hadoop-yarn-site/YARN.html` 39

Apache Storm, June 2019. `https://storm.apache.org/` 39

Y. Feunteun, Parallel and distributed deep learning: A survey, April 2019. `https://towardsdatascience.com/parallel-and-distributed-deep-learning-a-survey-97137ff94e4c` 33, 36

X. Tian, B. Xie, and J. Zhan, Cymbalo: An efficient graph processing framework for machine learning, In *Proc. of IEEE International Conference on Parallel and Distributed Processing*, December 2018. DOI: 10.1109/bdcloud.2018.00090. 39

Z. Zhang, P. Cui, and W. Zhu, Deep learning on graphs: A survey, *ArXiv Preprint ArXiv:1812.04202*, December 2018. `https://arxiv.org/abs/1812.04202` 39

W. Xiao, J. Xue, Y. Miao, et al., Tux^2: Distributed graph computation for machine learning, In *Proc. of the 14th USENIX Symposium on Networked Systems Design and Implementation (NSDI)*, March 2017. 39

Apache DeepSpark, June 2019. `http://deepspark.snu.ac.kr/` 35, 38

H. Kim, J. Park, J. Jang, et al., Deepspark: Spark-based deep learning supporting asynchronous updates and Caffe compatibility, *ArXiv Preprint ArXiv:1602.08191*, October 2016. `https://arxiv.org/abs/1602.08191` 38

Z. Jia, S. Lin, C. R. Qi, et al., Exploring hidden dimensions in accelerating conutional neural networks, In *Proc. of the 35th International Conference on Machine Learning (ICML)*, July 2018. `https://cs.stanford.edu/zhihao/papers/icml18full.pdf` 38

Z. Jia, M. Zaharia, and A. Aiken, Beyond data and model parallelism for deep neural networks, In *Proc. of the Conference on Systems and Machine Learning (SysML)*, April 2019. 37, 38

A. L. Gaunt, M. A. Johnson, A. Lawrence, et al., AMPNet: Asynchronous model-parallel training for dynamic neural networks, In *Proc. of the 6th International Conference on Learning Representations*, May 2018. 38

T. Chilimbi, Y. Suzue, J. Apacible, et al., Project Adam: Building an efficient and scalable deep learning training system, In *Proc. of the 11th USENIX Conference on Operating Systems Design and Implementation (OSDI)*, pp. 571–582, October 2014. 38

J. Dean, G. Corrado, R. Monga, et al., Large scale distributed deep networks, In *Proc. of the 25th International Conference on Neural Information Processing Systems (NIPS)*, pp. 1223–1231, December 2012. 10, 38, 40

K. Fukuda, Technologies behind distributed deep learning: AllReduce, July 2018. `https://preferredresearch.jp/2018/07/10/technologies-behind-distributed-deep-learning-allreduce/` 37

M. Li, D. G. Andersen, J. W. Park, et al., Scaling distributed machine learning with the parameter server, In *Proc. of the 11th USENIX Conference on Operating Systems Design and Implementation (OSDI)*, pp. 583–598, October 2014. DOI: 10.1145/2640087.2644155. 10, 33, 37, 52

A. Das, Distributed training of deep learning models with PyTorch, April 2019. `https://medium.com/intel-student-ambassadors/distributed-training-of-deep-learning-models-with-pytorch-1123fa538848` 37, 38

S. Wang, Distributed machine learning, January 2016. `https://www.slideshare.net/stanleywanguni/distributed-machine-learning?from_action=save` 10, 37, 38, 39

Google Inc., Distributed training in TensorFlow, June 2019. `https://www.tensorflow.org/guide/distribute_strategy` 35

S. Arnold, Writing distributed applications with PyTorch, June 2019. `https://pytorch.org/tutorials/intermediate/dist_tuto.html` 36

Microsoft, Distributed machine learning Toolkit (DMTK), June 2019. `http://www.dmtk.io/` 35

Turi-Create, June 2019. `https://turi.com/` 35

G. Malewicz, M. H. Austern, A. J. C. Bik, et al., Pregel: A system for large-scale graph processing, In *Proc. of the ACM SIGMOD International Conference on Management of Data (SIGMOD)*, June 2010. DOI: 10.1145/1807167.1807184. 35

Y. Low, J. Gonzalez, A. Kyrola, et al., GraphLab: A new framework for parallel machine learning, *ArXiv Preprint ArXiv:1006.4990*, June 2010. `https://arxiv.org/abs/1006.4990` 35, 39

Apache Spark MLlib, June 2019. `https://spark.apache.org/mllib/` 35

Apache Spark GraphX, June 2019. https://spark.apache.org/docs/latest/graphx-pr ogramming-guide.html 35

T. Ben-Nun and T. Hoefler, Demystifying parallel and distributed deep learning: An in-depth concurrency analysis, *ArXiv Preprint ArXiv:1802.09941*, September 2018. https://arxiv. org/abs/1802.09941 DOI: 10.1145/3320060. 10, 33, 36

J. Devlin, M. W. Chang, K. Lee, et al., BERT: Pre-training of deep bidirectional transformers for language understanding, *ArXiv Preprint ArXiv:1810.04805*, May 2019. https://arxi v.org/abs/1810.04805 34, 38

A. Galakatos, A. Crotty, and T. Kraska, Distributed machine learning, In *Encyclopedia of Database Systems*, December 2018. DOI: 10.1007/978-1-4614-8265-9_80647. 33, 36

R. Bekkerman, M. Bilenko, and J. Langford, *Scaling up machine learning: Parallel and distributed approaches*, Cambridge University Press, February 2012. DOI: 10.1017/cbo9781139042918. 33, 36

Y. Liu, J. Liu, and T. Basar, Differentially private gossip gradient descent, In *IEEE Conference on Decision and Control (CDC)*, pp. 2777–2782, December 2018. DOI: 10.1109/cdc.2018.8619437. 54

J. Daily, A. Vishnu, C. Siegel, et al., GossipGraD: Scalable deep learning using gossip communication based asynchronous gradient descent, *ArXiv Preprint ArXiv:1803.05880*, March 2018. http://arxiv.org/abs/1803.05880 54

C. Hardy, E. Le Merrer, and B. Sericola, Gossiping GANs: Position paper, In *Proc. of the 2nd Workshop on Distributed Infrastructures for Deep Learning*, pp. 25–28, December 2018. DOI: 10.1145/3286490.3286563. 54

I. Hegedüs, G. Danner, and M. Jelasity, Gossip learning as a decentralized alternative to federated learning, In *Proc. of the 14th International Federated Conference on Distributed Computing Techniques*, pp. 74–90, June 2019. DOI: 10.1007/978-3-030-22496-7_5. 54

D. Liu, T. Miller, R. Sayeed, et al., FADL: Federated-autonomous deep learning for distributed electronic health record, *ArXiv Preprint ArXiv:1811.11400*, November 2018. https://arxi v.org/abs/1811.11400 56

T. Nishio and R. Yonetani, Client selection for federated learning with heterogeneous resources in mobile edge, *ArXiv Preprint ArXiv:1804.08333*, October 2018. https://arxiv.org/ab s/1804.08333 DOI: 10.1109/icc.2019.8761315. 64

I. J. Goodfellow, O. Vinyals, and A. M. Saxe, Qualitatively characterizing neural network optimization problems, *ArXiv Preprint ArXiv:1412.6544*, May 2015. https://arxiv.org/ab s/1412.6544 60, 65

S. Ioffe and C. Szegedy, Batch normalization: Accelerating deep network training by reducing internal covariate shift, In *Proc. of the 32nd International Conference on Machine Learning (ICML)*, July 2015. 57

H. Tang, C. Yu, C. Renggli, et al., Distributed learning over unreliable networks, *ArXiv Preprint ArXiv:1810.07766*, May 2019. https://arxiv.org/abs/1810.07766 51, 52, 56, 60

Q. Ho, J. Cipar, H. Cui, et al., More effective distributed machine learning via a stale synchronous parallel parameter server, In *Proc. of the 26th International Conference on Neural Information Processing Systems (NIPS)*, pp. 1223–1231, December 2013. 52

H. Su and H. Chen, Experiments on parallel training of deep neural network using model averaging, *ArXiv Preprint ArXiv:1507.01239*, July 2018. https://arxiv.org/abs/1507.01239 51, 52, 60

X. Shu, G.-J. Qi, J. Tang, et al., Weakly-shared deep transfer networks for heterogeneous-domain knowledge propagation, In *Proc. of the 23rd ACM International Conference on Multimedia (MM)*, pp. 35–44, 2015. DOI: 10.1145/2733373.2806216. 87

F. Seide, G. Li, and D. Yu, Conversational speech transcription using context-dependent deep neural networks, In *12th Annual Conference of the International Speech Communication Association*, pp. 437–440, 2011.

M. Kamp, L. Adilova, J. Sicking, et al., Efficient decentralized deep learning by dynamic model averaging, In *Proc. of Machine Learning and Knowledge Discovery in Databases (KDD)*, pp. 393–409, September 2018. DOI: 10.1007/978-3-030-10925-7_24. 63

S. Han, H. Mao, and W. J. Dally, Deep compression: Compressing deep neural networks with pruning, trained quantization and Huffman coding, *ArXiv Preprint ArXiv:1510.00149*, February 2016. https://arxiv.org/abs/1510.00149 63

Y. Lin, S. Han, H. Mao, et al., Deep gradient compression: Reducing the communication bandwidth for distributed training, In *International Conference on Learning Representations (ICLR)*, April 2018. 50

E. Zhong, W. Fan, Q. Yang, et al., Cross validation framework to choose amongst models and datasets for transfer learning, In J. L. Balcázar, F. Bonchi, A. Gionis, and M. Sebag, Eds., *Machine Learning and Knowledge Discovery in Databases*, pp. 547–562, Springer, Heidelberg, 2010. DOI: 10.1007/978-3-642-15883-4.

I. Kuzborskij and F. Orabona, Stability and hypothesis transfer learning, In *Proc. of the 30th International Conference on Machine Learning (ICML)*, 28(3):942–950, 2013.

B. Hitaj, G. Ateniese, and F. Pérez-Cruz, Deep models under the GAN: Information leakage from collaborative deep learning, In *Proc. of the ACM SIGSAC Conference on Computer and Communications Security*, pp. 603–618, October 2017. DOI: 10.1145/3133956.3134012. 10, 50, 52, 89

F. McSherry, Deep learning and differential privacy, `https://github.com/frankmcsherry/blog/blob/master/posts/2017-10-27.md` 89

Z. Li, Y. Zhang, Y. Wei, et al., End-to-end adversarial memory network for cross-domain sentiment classification, In *Proc. of the 26th International Joint Conference on Artificial Intelligence (IJCAI)*, pp. 2237–2243, August 2017. DOI: 10.24963/ijcai.2017/311. 84

S. J. Pan, X. Ni, J.-T. Sun, et al., Cross-domain sentiment classification via spectral feature alignment, In *Proc. of the 19th International Conference on World Wide Web (WWW)*, pp. 751–760, April 2010. DOI: 10.1145/1772690.1772767. 84

Y. Zhu, Y. Chen, Z. Lu, et al., Heterogeneous transfer learning for image classification, In *Proc. of the 25th AAAI Conference on Artificial Intelligence (AAAI)*, pp. 1304–1309, August 2011. 84

M. Oquab, L. Bottou, I. Laptev, et al., Learning and transferring mid-level image representations using conutional neural networks, In *Proc. of the IEEE Conference on Computer Vision and Pattern Recognition CVPR*, pp. 1717–1724, June 2014. DOI: 10.1109/cvpr.2014.222. 86

R. Bahmani, M. Barbosa, F. Brasser, et al., Secure multiparty computation from SGX, In *Proc. of International Conference on Financial Cryptography and Data Security Financial Cryptography and Data Security (FC)*, pp. 477–497, December 2017. DOI: 10.1007/978-3-319-70972-7_27. 71

T. Chen and C. Guestrin, XGBoost: A scalable tree boosting system, In *Proc. of the 22nd International Conference on Knowledge Discovery and Data Mining (KDD)*, pp. 785–794, August 2016. DOI: 10.1145/2939672.2939785. 77, 78

K. Chang, N. Balachandar, C. K. Lam, et al., Institutionally distributed deep learning networks, *ArXiv Preprint ArXiv:1709.05929*, September 2017. `https://arxiv.org/abs/1709.05929` 53

K. Chang, N. Balachandar, C. Lam, et al., Distributed deep learning networks among institutions for medical imaging, *Journal of the American Medical Informatics Association*, 25(8):945–954, August 2018. DOI: 10.1093/jamia/ocy017. 53

L. T. Phong, Privacy-preserving stochastic gradient descent with multiple distributed trainers, In *Proc. of the 11th International Conference on Network and System Security (NSS)*, pp. 510–518, July 2017. DOI: 10.1007/978-3-319-64701-2_38. 10

L. T. Phong and T. T. Phuong, Privacy-preserving deep learning via weight transmission, *IEEE Transactions on Information Forensics and Security*, April 2019. `https://arxiv.org/abs/1809.03272` DOI: 10.1109/tifs.2019.2911169. 33, 52, 53

L. T. Phong, Y. Aono, T. Hayashi, et al., Privacy-preserving deep learning via additively homomorphic encryption, *IEEE Transactions on Information Forensics and Security*, 13(5):1333–1345, May 2018. DOI: 10.1109/tifs.2017.2787987. 10, 49, 50, 51, 52, 60, 62, 88, 89

L. Su and J. Xu, Securing distributed gradient descent in high dimensional statistical learning, In *Proc. of the ACM on Measurement and Analysis of Computing Systems*, 3(1), March 2019. DOI: 10.1145/3309697.3331499.

S. Tutdere and O. Uzunko, Construction of arithmetic secret sharing schemes by using torsion limits, *ArXiv Preprint ArXiv:1506.06807*, June 2015. `https://arxiv.org/abs/1506.06807` DOI: 10.15672/hujms.460348. 24

A. Beimel, Secret-sharing schemes: A Survey, *IWCC*, LNCS 6639, pp. 11–46, Springer-Verlag, 2011. DOI: 10.1007/978-3-642-20901-7_2. 24

A. Acar, H. Aksu, A. S. Uluagac, et al., A survey on homomorphic encryption schemes: Theory and implementation, *ACM Computing Surveys (CSUR)*, 51(4):79:1–79:35, 2018. DOI: 10.1145/3214303. 27, 28, 62, 88

Y. Aono, T. Hayashi, L. Wang, et al., Privacy-preserving deep learning via additively homomorphic encryption, *IEEE Transactions on Information Forensics and Security*, 13(5):1333–1345, 2018. DOI: 10.1109/TIFS.2017.2787987. 19, 46

F. Armknecht, C. Boyd, C. Carr, et al., A guide to fully homomorphic encryption, *IACR Cryptology ePrint Archive*, 2015. `https://eprint.iacr.org/2015/1192.pdf` 27

H. Bae, D. Jung, and S. Yoon, Anomigan: Generative adversarial networks for anonymizing private medical data, *ArXiv Preprint ArXiv:1901.11313*, January 2019. `https://arxiv.org/abs/1901.11313`

E. Bagdasaryan, A. Veit, Y. Hua, et al., How to backdoor federated learning, *ArXiv Preprint ArXiv:1807.00459*, August 2019. `https://arxiv.org/abs/1807.00459` 20, 140

M. Barreno, B. Nelson, R. Sears, et al., Can machine learning be secure? In *Proc. of the ACM Symposium on Information, Computer and Communications Security*, pp. 16–25, 2006. DOI: 10.1145/1128817.1128824. 17, 18

D. Beaver, Efficient multiparty protocols using circuit randomization, In *Proc. of the Annual International Cryptology Conference*, pp. 420–432, Springer, 1991. DOI: 10.1007/3-540-46766-1_34. 24, 48

D. Beaver, Correlated pseudorandomness and the complexity of private computations, In *Proc. STOC Proceedings of the 28th Annual ACM Symposium on Theory of Computing*, pp. 479–488, May 1996. DOI: 10.1145/237814.237996. 23

M. Bellare and S. Micali, Non-interactive oblivious transfer and applications, In G. Brassard, Ed., *Advances in Cryptology—CRYPTO Proceedings*, pp. 547–557, Springer, New York, 1990. DOI: 10.1007/0-387-34805-0. 22

M. Ben-or, S. Goldwasser, and A. Wigderson, Completeness theorems for non-cryptographic fault-tolerant distributed computation (extended abstract), In *Proc. of the 20th Annual ACM Symposium on Theory of Computing*, pp. 1–10, January 1988. DOI: 10.1145/62212.62213.

D. Bogdanov, L. Kamm, S. Laur, et al., Privacy-preserving statistical data analysis on federated databases, In *Annual Privacy Forum*, pp. 30–55, Springer, 2014. DOI: 10.1007/978-3-319-06749-0_3. 10, 18

D. Boneh, R. Gennaro, S. Goldfeder, et al., Threshold cryptosystems from threshold fully homomorphic encryption, In H. Shacham and A. Boldyreva, Eds., *Advances in Cryptology—CRYPTO*, pp. 565–596, Springer International Publishing, 2018. DOI: 10.1007/978-3-319-96881-0.

D. Boneh, E.-J. Goh, and K. Nissim, Evaluating 2-DNF formulas on ciphertexts, In *Theory of Cryptography Conference*, pp. 325–341, Springer, 2005. DOI: 10.1007/978-3-540-30576-7_18. 26, 28

K. Bonawitz, V. Ivanov, B. Kreuter, et al., Practical secure aggregation for privacy-preserving machine learning, In *Proc. of the ACM SIGSAC Conference on Computer and Communications Security (CCS)*, pp. 1175–1191, November 2017. DOI: 10.1145/3133956.3133982. 25, 49, 51, 52

K. Bonawitz, H. Eichner, W. Grieskamp, et al., Towards federated learning at scale: System design, *ArXiv Preprint ArXiv:1902.01046*, March 2019. https://arxiv.org/abs/1902.01046 11, 56, 65, 66, 98, 112, 141, 143

Z. Brakerski, C. Gentry, and V. Vaikuntanathan, Fully homomorphic encryption without bootstrapping, *IACR Cryptology ePrint Archive*, 2011. 28

Z. Brakerski and V. Vaikuntanathan, Fully homomorphic encryption from ring-LWE and security for key dependent messages, In P. Rogaway, Ed., *Advances in Cryptology—CRYPTO*, pp. 505–524, Springer, 2011. DOI: 10.1007/978-3-642-22792-9.

R. Canetti, Universally composable security: A new paradigm for cryptographic protocols, In *Proc. IEEE International Conference on Cluster Computing*, pp. 136–145, October 2001. DOI: 10.1109/sfcs.2001.959888.

K. Chaudhuri, C. Monteleoni, and A. D. Sarwate, Differentially private empirical risk minimization, *Journal of Machine Learning Research*, pp. 1069–1109, March 2011.

D. Cozzo and N. P. Smart, Using TopGear in overdrive: A more efficient ZKPoK for SPDZ, *Cryptology ePrint Archive, Report 2019/035*, 2019. https://eprint.iacr.org/2019/035

I. Damård, V. Pastro, N. P. Smart, et al., Multiparty computation from somewhat homomorphic encryption, *Cryptology ePrint Archive, Report 2011/535*, 2011. https://eprint.iacr.org/2011/535 DOI: 10.1007/978-3-642-32009-5_38. 24, 25, 48

I. Damård, M. Keller, E. Larraia, et al., Practical covertly secure MPC for dishonest majority—or: Breaking the SPDZ limits, *Cryptology ePrint Archive, Report 2012/642*, 2012. https://eprint.iacr.org/2012/642 DOI: 10.1007/978-3-642-40203-6_1. 25

I. Damgård and J. B. Nielsen, Universally composable efficient multiparty computation from threshold homomorphic encryption, In D. Boneh, Ed., *Advances in Cryptology—CRYPTO*, pp. 247–264, Springer, 2003. DOI: 10.1007/978-3-540-45146-4_15. 22

D. Demmler, T. Schneider, and M. Zohner, Aby-a framework for efficient mixed-protocol secure two-party computation, In *NDSS*, February 2015. DOI: 10.14722/ndss.2015.23113. 23, 25

W. Diffie and M. E. Hellman, New directions in cryptography, *IEEE Transactions on Information Theory*, 22(6):644–654, November 1976. DOI: 10.1109/tit.1976.1055638. 23

N. Dowlin, R. Gilad-Bachrach, K. Laine, et al., CryptoNets: Applying neural networks to encrypted data with high throughput and accuracy, In *International Conference on Machine Learning*, pp. 201–210, June 2016.

W. Du, Y. S. Han, and S. Chen, Privacy-preserving multivariate statistical analysis: Linear regression and classification, In *Proc. of the SIAM International Conference on Data Mining*, pp. 222–233, Society for Industrial and Applied Mathematics, April 2004. DOI: 10.1137/1.9781611972740.21. 46, 75, 85

C. Dwork, Differential privacy, *Encyclopedia of Cryptography and Security*, pp. 338–340, 2011. DOI: 10.1007/978-1-4419-5906-5_752.

C. Dwork, K. Kenthapadi, F. McSherry, et al., Our data, ourselves: Privacy via distributed noise generation, In *Annual International Conference on the Theory and Applications of Cryptographic Techniques*, pp. 486–503, Springer, 2006. DOI: 10.1007/11761679_29. 30

C. Dwork, V. Feldman, M. Hardt, et al., Preserving statistical validity in adaptive data analysis, *ArXiv Preprint ArXiv:1411.2664*, March 2016. https://arxiv.org/abs/1411.2664 29

C. Dwork, F. McSherry, K. Nissim, et al., Calibrating noise to sensitivity in private data analysis, In *Theory of Cryptography Conference*, pp. 265–284, Springer, 2006. DOI: 10.1007/11681878_14. 20, 29, 30

C. Dwork and K. Nissim, Privacy-preserving data mining on vertically partitioned databases, In *Annual International Cryptology Conference*, Springer, pp. 528–544, 2004. DOI: 10.1007/978-3-540-28628-8_32. 30

T. ElGamal, A public key cryptosystem and a signature scheme based on discrete logarithms, *IEEE Transactions on Information Theory*, 31(4):469–472, 1985. DOI: 10.1007/3-540-39568-7_2. 27

C. Fontaine and F. Galand, A survey of homomorphic encryption for nonspecialists, *EURASIP Journal on Information Security*, (15), 2007. DOI: 10.1186/1687-417x-2007-013801.

A. Gascón, P. Schoppmann, B. Balle, et al., Secure linear regression on vertically partitioned datasets, *IACR Cryptology ePrint Archive*, 2016.

C. Gentry, Fully homomorphic encryption using ideal lattices, In *Proc. of the 41st Annual ACM Symposium on Theory of Computing*, 9:169–178, June 2009. DOI: 10.1145/1536414.1536440. 26, 28

N. Gilboa, Two party RSA key generation, In *Annual International Cryptology Conference*, pp. 116–129, Springer, 1999. DOI: 10.1007/3-540-48405-1_8. 24

O. Goldreich, S. Micali, and A. Wigderson, How to play any mental game, In *Proc. of the 19th Annual ACM Symposium on Theory of Computing*, pp. 218–229, January 1987. DOI: 10.1145/28395.28420. 22, 23, 25, 71

S. Goldwasser and S. Micali, Probabilistic encryption and how to play mental poker keeping secret all partial information, In *Proc. of the 14th Annual ACM Symposium on Theory of Computing*, pp. 365–377, 1982. DOI: 10.1145/800070.802212. 26, 27

T. Gu, B. Dolan-Gavitt, and S. Garg, Badnets: Identifying vulnerabilities in the machine learning model supply chain, *ArXiv Preprint ArXiv:1708.06733*, August 2017. https://arxiv.org/abs/1708.06733

S. Hardy, W. Henecka, H. Ivey-Law, et al., Private federated learning on vertically partitioned data via entity resolution and additively homomorphic encryption, *ArXiv Preprint ArXiv:1711.10677*, November 2017. https://arxiv.org/abs/1711.10677 28

C. Hazay and Y. Lindell, Efficient secure two-party protocols, In *Information Security and Cryptography*, 2010. DOI: 10.1007/978-3-642-14303-8. 22

X. He, A. Prasad, S. P. Sethi, et al., A survey of Stackelberg differential game models in supply and marketing channels, *Journal of Systems Science and Systems Engineering*, 16:385–413, 2007. DOI: 10.1007/s11518-008-5082-x.

R. Impagliazzo and S. Rudich, Limits on the provable consequences of one-way permutations, In *Proc. of the 21st Annual ACM Symposium on Theory of Computing (STOC)*, pp. 44–61, 1989. DOI: 10.1145/73007.73012. 23

Y. Ishai and A. Paskin, Evaluating branching programs on encrypted data, In S.P. Vadhan, Ed., *Theory of Cryptography*, pp. 575–594, Springer, 2007. DOI: 10.1007/978-3-540-70936-7. 28

Y. Ishai, M. Prabhakaran, and A. Sahai, Founding cryptography on oblivious transfer—efficiently, In David Wagner, Ed., *Advances in Cryptology—CRYPTO*, pp. 572–591, Springer Berlin Heidelberg, Berlin, Heidelberg, 2008. DOI: 10.1007/978-3-540-85174-5. 22

B. Jayaraman and D. Evans, When relaxations go bad: Differentially-private machine learning, *ArXiv Preprint ArXiv:1902.08874*, February 2019. https://arxiv.org/abs/1902.08874 30

L. Jiang, R. Tan, X. Lou, et al., On lightweight privacy-preserving collaborative learning for internet-of-things objects, In *IoTDI*, 2019. DOI: 10.1145/3302505.3310070. 47

A. F. Karr, X. S. Lin, A. P. Sanil, et al., Privacy-preserving analysis of vertically partitioned data using secure matrix products, *Journal of Official Statistics*, pp. 125–138, September 2009.

M. Kim, Y. Song, S. Wang, et al., Secure logistic regression based on homomorphic encryption: Design and evaluation, *JMIR Medical Informatics*, 6(2), April 2018. DOI: 10.2196/preprints.8805. 88

Y. Lindell, Secure multiparty computation for privacy preserving data mining, In *Encyclopedia of Data Warehousing and Mining*, pp. 1005–1009, IGI Global, 2005. DOI: 10.4018/9781591405573.ch189. 21

Y. Lindell, How to simulate it—a tutorial on the simulation proof technique, In Y. Lindell, Ed., *Tutorials on the Foundations of Cryptography, Information Security and Cryptography*, pp. 277–346, Springer, April 2017. DOI: 10.1007/978-3-319-57048-8. 22

Y. Lindell and B. Pinkas, Secure multiparty computation for privacy-preserving data mining, *IACR Cryptology ePrint Archive*, 1(1):59–98, April 2009. DOI: 10.4018/9781591405573.ch189. 21

A. López-Alt, E. Tromer, and V. Vaikuntanathan, On-the-fly multiparty computation on the cloud via multi-key fully homomorphic encryption, In *Proc. of the 44th Annual ACM Symposium on Theory of Computing*, pp. 1219–1234, 2012. DOI: 10.1145/2213977.2214086. 28

V. Lyubashevsky, C. Peikert, and O. Regev, On ideal lattices and learning with errors over rings, In *Annual International Conference on the Theory and Applications of Cryptographic Techniques*, pp. 1–23, Springer, 2010. DOI: 10.1145/2535925. 28

F. McSherry and K. Talwar, Mechanism design via differential privacy, In *FOCS*, 7:94–103, 2007. DOI: 10.1109/focs.2007.66. 30

D. Mishra and D. Veeramani, Vickrey–Dutch procurement auction for multiple items, *European Journal of Operational Research*, 180:617–629, 2007. DOI: 10.1016/j.ejor.2006.04.020. 99

P. Mohassel and P. Rindal, ABY3: A mixed protocol framework for machine learning, In *Proc. of the ACM SIGSAC Conference on Computer and Communications Security CCS*, pp. 35–52, October 2018. DOI: 10.1145/3243734.3243760.

M. Naor and B. Pinkas, Efficient oblivious transfer protocols, In *Proc. of the 12th Annual ACM-SIAM Symposium on Discrete Algorithms*, pp. 448–457, Society for Industrial and Applied Mathematics, January 2001. 22

A. Narayanan and V. Shmatikov, Robust de-anonymization of large datasets (how to break anonymity of the Netflix prize dataset), University of Texas at Austin, February 2008. 20

P. Paillier, Public-key cryptosystems based on composite degree residuosity classes, In *International Conference on the Theory and Applications of Cryptographic Techniques*, pp. 223–238, Springer, Berlin, Heidelberg, May 1999. DOI: 10.1007/3-540-48910-x_16. 26, 27, 61, 62, 78

N. Phan, Y. Wang, X. Wu, et al., Differential privacy preservation for deep auto-encoders: an application of human behavior prediction, In *30th AAAI Conference on Artificial Intelligence*, February 2016.

M. O. Rabin, How to exchange secrets with oblivious transfer, Harvard University Technical Report, May 1981. https://eprint.iacr.org/2005/187.pdf 22

T. Rabin and M. Ben-Or, Verifiable secret sharing and multiparty protocols with honest majority, In *Proc. of the 21st Annual ACM Symposium on Theory of Computing STOC*, pp. 73–85, New York, 1989. DOI: 10.1145/73007.73014. 22

L. Reyzin, A. D. Smith, and S. Yakoubov, Turning HATE into LOVE: Homomorphic ad hoc threshold encryption for scalable MPC, *IACR Cryptology ePrint Archive*, 2018.

R. L. Rivest, A. Shamir, and L. Adleman, A method for obtaining digital signatures and public-key cryptosystems, *Communications of the ACM*, 21(2):120–126, 1978. DOI: 10.21236/ada606588. 27

B. D. Rouhani, M. S. Riazi, and F. Koushanfar, DeepSecure: Scalable provably-secure deep learning, *ArXiv Preprint ArXiv:1705.08963*, May, 2017. https://arxiv.org/abs/1705.08963 DOI: 10.1109/dac.2018.8465894. 25

P. Samarati and L. Sweeney, Protecting privacy when disclosing information: k-anonymity and its enforcement through generalization and suppression, *Technical Report*, SRI International, 1998.

A. P. Sanil, A. F. Karr, X. Lin, et al., Privacy preserving regression modelling via distributed computation, In *Proc. of the 10th ACM SIGKDD International Conference on Knowledge Discovery and Data Mining*, pp. 677–682, August 2004. DOI: 10.1145/1014052.1014139.

N. P. Smart, The discrete logarithm problem on elliptic curves of trace one, *Journal of Cryptology*, 12:193–196, 1999. DOI: 10.1007/s001459900052.

J. Vaidya and C. Clifton, Privacy preserving association rule mining in vertically partitioned data, In *Proc. of the 8th ACM SIGKDD International Conference on Knowledge Discovery and Data Mining*, pp. 639–644, July 2002. DOI: 10.1145/775047.775142. 75

M. V. Dijk, C. Gentry, S. Halevi, et al., Fully homomorphic encryption over the integers, In *Annual International Conference on the Theory and Applications of Cryptographic Techniques*, pp. 24–43, Springer, 2010. DOI: 10.1007/978-3-642-13190-5_2. 28

S. Wagh, D. Gupta, and N. Chandran, SecureNN: Efficient and private neural network training, *IACR Cryptology ePrint Archive*, 2018. 48

Z. Brakerski and V. Vaikuntanathan, Efficient fully homomorphic encryption from (standard) LWE, In *IEEE 52nd Annual Symposium on Foundations of Computer Science*, pp. 97–106, October 2011. DOI: 10.1109/focs.2011.12. 28

L. Wan, W. K. Ng, S. Han, et al., Privacy-preservation for gradient descent methods, In *Proc. of the 13th ACM SIGKDD International Conference on Knowledge Discovery and Data Mining*, pp. 775–783, August 2007. DOI: 10.1145/1281192.1281275. 45, 46

J.-S. Weng, J. Weng, M. Li, et al., DeepChain: Auditable and privacy-preserving deep learning with blockchain-based incentive, *IACR Cryptology ePrint Archive*, 2018.

A. F. Westin, Privacy and freedom, Washington Lee Law Review, 1968. 17

C. Xie, S. Koyejo, and I. Gupta, SLSGD: Secure and efficient distributed on-device machine learning, *ArXiv Preprint ArXiv:1903.06996*, March 2019. https://arxiv.org/abs/1903.06996

Y. Yin, I. Kaku, J. Tang, et al., *Application for privacy-preserving data mining*, Springer London, London, 2011. DOI: 10.1007/978-1-84996-338-1_14.

M. Chen, R. Mathews, T. Ouyang, et al., Federated learning of out-of-vocabulary words, *ArXiv Preprint ArXiv:1903.10635*, March 2019. https://arxiv.org/abs/1903.10635 11, 112

A. Sergeev and M. D. Balso, Horovod: Fast and easy distributed deep learning in TensorFlow, *ArXiv Preprint ArXiv:1802.05799*, February 2018. https://arxiv.org/abs/1802.05799 12

coMind.org, Machine learning network for deep learning, https://comind.org/ 12

V. Smith, C.-K. Chiang, M. Sanjabi, et al., Federated multi-task learning, In *Proc. of International Conference on Neural Information Processing Systems (NIPS)*, December 2017. 11, 50

H. H. Zhuo, W. Feng, Q. Xu, et al., Federated reinforcement learning, *ArXiv Preprint ArXiv:1901.08277*, January 2019. https://arxiv.org/abs/1901.08277 11, 126, 130

WeBank AI Department, Federated learning white paper V1.0, September 2018. https://aisp-1251170195.cos.ap-hongkong.myqcloud.com/fedweb/1552917186945.pdf 2, 5

S. Pouyanfar, S. Sadiq, Y. Yan, et al., A survey on deep learning: Algorithms, techniques, and applications, *ACM Computing Survey*, 51(5):1–36, January 2019. DOI: 10.1145/3234150. 1

W. G. Hatcher and W. Yu, A survey of deep learning: Platforms, applications and emerging research trends, *IEEE Access*, 6(24):411–432, April 2018. DOI: 10.1109/access.2018.2830661. 1

I. Goodfellow, Y. Bengio and A. Courville, *Deep Learning*, MIT Press, April 2016. http://www.deeplearningbook.org 1, 57

The official GDPR website https://ec.europa.eu/commission/priorities/justice-and-fundamental-rights/data-protection/2018-reform-eu-data-protection-rules_en 2

DLA Piper, Data protection laws of the world: Full handbook, September 2019. https://www.dlapiperdataprotection.com/ 2, 143, 147, 151, 152, 153

Q. Yang, Y. Liu, T. Chen, et al., Federated machine learning: Concept and applications, *ArXiv Preprint ArXiv:1902.04885*, February 2019. http://arxiv.org/abs/1902.04885 DOI: 10.1145/3298981. xiv, 2, 3, 5, 7, 8, 9, 10, 34, 50, 51, 66, 70, 71, 72, 73, 83, 85, 86, 151, 153

H. B. McMahan, E. Moore, D. Ramage, et al., Communication-efficient learning of deep networks from decentralized data, *ArXiv Preprint ArXiv:1602.05629*, February 2016a. https://arxiv.org/abs/1602.05629 3, 5, 12, 49, 50, 51, 52, 55, 56, 60, 69, 85

H. B. McMahan, E. Moore, D. Ramage, et al., Federated learning of deep networks using model averaging, February 2016b. `https://pdfs.semanticscholar.org/8b41/9080c d37bdc30872b76f405ef6a93eae3304.pdf` 3, 44, 45, 55, 56, 57, 58, 59, 60, 64, 118

H. Yu, S. Yang, and S. Zhu, Parallel restarted SGD with faster convergence and less communication: Demystifying why model averaging works for deep learning, *ArXiv Preprint ArXiv:1807.06629*, July 2018. `https://arxiv.org/abs/1807.06629` DOI: 10.1609/aaai.v33i01.33015693. 12, 33, 52, 55, 60

J. Konecný, H. B. McMahan, F. X. Yu, et al., Federated learning: Strategies for improving communication efficiency, *ArXiv Preprint ArXiv:1610.05492*, October 2016a. `http://arxi v.org/abs/1610.05492` 3, 47, 63

J. Konecný, H. B. McMahan, D. Ramage, et al., Federated optimization: Distributed machine learning for on-device intelligence, *ArXiv Preprint ArXiv:1610.02527*, October 2016b. `http: //arxiv.org/abs/1610.02527` 3, 33, 50, 65

F. Hartmann, Federated learning, Master thesis, Free University of Berlin, May 2018. `http: //www.mi.fu-berlin.de/inf/groups/ag-ti/theses/download/Hartmann_F18.pdf` 3, 6, 11, 66

F. Hartmann, Federated learning, August 2018. `https://florian.github.io/federated- learning/` 1, 56, 66

Y. Liu, Q. Yang, T. Chen, et al., Federated learning and transfer learning for privacy, security and confidentiality, *The 33rd AAAI Conference on Artificial Intelligence (AAAI)*, January 2019. `https://aisp-1251170195.file.myqcloud.com/fedweb/1552916850679.pdf` 3, 5, 7, 71, 83

T. Yang, G. Andrew, H. Eichner, et al., Applied federated learning: Improving Google keyboard query suggestions, *ArXiv Preprint ArXiv:1812.02903*, December 2018. `http://arxiv.org/ abs/1812.02903` 3, 11, 65

A. Hard, K. Rao, R. Mathews, et al., Federated learning for mobile keyboard prediction, *ArXiv Preprint ArXiv:1811.03604*, November 2018. `http://arxiv.org/abs/1811.03604` 3, 11, 65

Y. Zhao, M. Li, L. Lai, et al., Federated learning with non-IID data, *ArXiv Preprint ArXiv:1806.00582*, August 2018. `http://arxiv.org/abs/1806.00582` 6, 57

F. Sattler, S. Wiedemann, K. Müller, et al., Robust and communication-efficient federated learning from non-IID data, *ArXiv Preprint ArXiv:1903.02891*, March 2019. `http://ar xiv.org/abs/1903.02891` 6, 56, 57

S. van Lier, Robustness of federated averaging for non-IID data, August 2018. `https://www.cs.ru.nl/bachelors-theses/2018/Stan_van_Lier___4256166___Robustness_of_federated_averaging_for_non-IID_data.pdf` 6

A. N. Bhagoji, S. Chakraborty, P. Mittal, et al., Analyzing federated learning through an adversarial lens, *ArXiv Preprint ArXiv:1811.12470*, March 2019. `http://arxiv.org/abs/1811.12470` 7, 11, 20

B. Han, An overview of federated learning, March 2019. `https://medium.com/datadriveninvestor/an-overview-of-federated-learning-8a1a62b0600d` 7, 12, 13

J. Mancuso, B. DeCoste, and G. Uhma, Privacy-preserving machine learning 2018: A year in review, January 2019. `https://medium.com/dropoutlabs/privacy-preserving-machine-learning-2018-a-year-in-review-b6345a95ae0f` 10, 17, 34, 145, 153

K. Cheng, T. Fan, Y. Jin, et al., Secureboost: A lossless federated learning framework, *ArXiv Preprint ArXiv:1901.08755*, January 2019. `http://arxiv.org/abs/1901.08755` 10, 42, 72, 76, 78, 79

Y. Liu, T. Chen, and Q. Yang, Secure federated transfer learning, *ArXiv Preprint ArXiv:1812.03337*, January 2018. `http://arxiv.org/abs/1812.03337` 10, 29, 48, 83, 85, 86, 89, 90

F. Chen, Z. Dong, Z. Li, et al., Federated meta-learning for recommendation, *ArXiv Preprint ArXiv:1802.07876*, February 2018. `http://arxiv.org/abs/1802.07876` 11, 118

D. Liu, T. Miller, R. Sayeed, et al., FADL: Federated-autonomous deep learning for distributed electronic health record, *ArXiv Preprint ArXiv:1811.11400*, November 2018. `http://arxiv.org/abs/1811.11400` 11

L. Huang and D. Liu, Patient clustering improves efficiency of federated machine learning to predict mortality and hospital stay time using distributed electronic medical records, *ArXiv Preprint ArXiv:1903.09296*, March 2019. `http://arxiv.org/abs/1903.09296` DOI: 10.1016/j.jbi.2019.103291. 11

OpenMined. `https://www.openmined.org/` 13

Horovod. `https://github.com/horovod` 12

T. Ryffel, A. Trask, M. Dahl, et al., A generic framework for privacy preserving deep learning, *ArXiv Preprint ArXiv:1811.04017*, November 2018. `http://arxiv.org/abs/1811.04017` 13

OpenMined/PySyft. `https://github.com/OpenMined/PySyft` 13, 144

T. Ryffel, Federated learning with PySyft and PyTorch, March 2019. `https://blog.openmin ed.org/upgrade-to-federated-learning-in-10-lines/` 13

WeBank AI Department, Federated AI technology enabler (FATE). `https://github.com/F ederatedAI/FATE` xv, 12, 14, 144

WeBank AI Department, The federated AI ecosystem project. `https://www.fedai.org/#/` 12, 14, 143

Tensorflow.org, Tensorflow federated (TFF): Machine learning on decentralized data. `https://www.tensorflow.org/federated` 12, 64

A. Ingerman and K. Ostrowski, Introducing tensorflow federated, March 2019. `https://medi um.com/tensorflow/introducing-tensorflow-federated-a4147aa20041` 12

Tensorflow/federated. `https://github.com/tensorflow/federated` 12

TensorFlow/Encrypted. `https://github.com/tf-encrypted/tf-encrypted` 12

coMindOrg/federated-averaging-tutorials. `https://github.com/coMindOrg/federated-averaging-tutorials` 12

IEEE P3652.1—Guide for architectural framework and application of federated machine learning. `https://standards.ieee.org/project/3652_1.html` and `https://sagroups.iee e.org/3652-1/` xv, 14

The general data protection regulation (GDPR), April 2016. `https://eur-lex.europa.eu/ legal-content/EN/TXT/HTML/?uri=CELEX:32016R0679&from=EN` 146, 147, 148, 149

GDPR Info. `https://gdpr-info.eu/` 145, 148, 149

EU GDPR.ORG. `https://eugdpr.org/` 145

Overview of the general data protection regulation (GDPR). `https://ico.org.uk/media /for-organisations/data-protection-reform/overview-of-the-gdpr-1-13.pdf` 145, 148, 149

TechRepublic, GDPR: A cheat sheet. `https://www.techrepublic.com/article/the-eu- general-data-protection-regulation-gdpr-the-smart-persons-guide/` 147

A. Kotsios, M. Magnani, L. Rossi, et al., An analysis of the consequences of the general data protection regulation (GDPR) on social network research, *ArXiv Preprint ArXiv:1903.03196*, March 2019. `http://arxiv.org/abs/1903.03196` 148, 149

A. Shah, V. Banakar, S. Shastri, et al., Analyzing the impact of GDPR on storage systems, *ArXiv Preprint ArXiv:1903.04880*, March 2019. `http://arxiv.org/abs/1903.04880` 154

A. Dasgupta and A. Ghosh, Crowdsourced judgement elicitation with endogenous proficiency, In *WWW*, pp. 319–330, 2013. DOI: 10.1145/2488388.2488417. 97

B. Faltings and G. Radanovic, *Game Theory for Data Science: Eliciting Truthful Information*, Morgan & Claypool Publishers, 2017. DOI: 10.2200/s00788ed1v01y201707aim035. 97

R. Jia, D. Dao, B. Wang, F. A. Hubis, N. Hynes, N. M. Gurel, B. Li, C. Zhang, D. Song and C. Spanos, Towards efficient data valuation based on the Shapley value, In *PLMR*, pp. 1167–1176, 2019. 98

Y. Kong and G. Schoenebeck, An information theoretic framework for designing information elicitation mechanisms that reward truth-telling, *ACM Transactions on Economics and Computation*, 7(1), article 2, 2019. DOI: 10.1145/3296670. 97

G. Radanovic, B. Faltings and R. Jurca, Incentives for effort in crowdsourcing using the peer truth serum, *ACM Transactions on Intelligent Systems and Technology*, 7(4), article 48, 2016. DOI: 10.1145/2856102. 98

A. Richardson, A. Filos-Ratsikas and B. Faltings, Rewarding high-quality data via influence functions, *arXiv 1908.11598*, 2019. 98

A. Singla and A. Krause, Truthful incentives in crowdsourcing tasks using regret minimization mechanisms, In *WWW*, pp. 1167–1178, 2013. DOI: 10.1145/2488388.2488490. 97

V. Shnayder, A. Agarwal, R. Frongillo, and D. C. Parkes, Informed truthfulness in multi-task peer prediction, In *ACM EC*, pp. 179–196, 2016. DOI: 10.1145/2940716.2940790. 97

S. Sirur, J. R. C. Nurse, and H. Webb, Are we there yet? understanding the challenges faced in complying with the general data protection regulation (GDPR), *ArXiv Preprint ArXiv:1808.07338*, September 2018. http://arxiv.org/abs/1808.07338 DOI: 10.1145/3267357.3267368.

University of Groningen, Understanding the GDPR. https://www.futurelearn.com/courses/general-data-protection-regulation/0/steps/32412 148

T. McGavisk, The positive and negative impact of GDPR. https://www.timedatasecurity.com/blogs/the-positive-and-negative-implications-of-gdpr 150, 151

D. Roe, Understanding GDPR and its impact on the development of AI, April 2018. https://www.cmswire.com/information-management/understanding-gdpr-and-its-impact-on-the-development-of-ai/ 151

J. Pierce, Privacy and cybersecurity: A global year-end review, December 2018. https://www.insideprivacy.com/data-privacy/privacy-and-cybersecurity-a-global-year-end-review/ 151, 153

The California consumer privacy act (CCPA). https://www.caprivacy.org/ DOI: 10.2307/j.ctvjghvnn. 152

Information security technology—Personal information security specification. http://www.gb688.cn/bzgk/gb/newGbInfo?hcno=4FFAA51D63BA21B9EE40C51DD3CC40BE 153

G. Liang and S. S. Chawathe, Privacy-preserving inter-database operations, In *International Conference on Intelligence and Security Informatics*, pp. 66–82, Springer, 2004. DOI: 10.1007/978-3-540-25952-7_6. 71, 76

M. Scannapieco, I. Figotin, E. Bertino, et al., Privacy preserving schema and data matching, In *Proc. of the ACM SIGMOD International Conference on Management of Data*, pp. 653–664, 2007. DOI: 10.1145/1247480.1247553. 71

R. Nock, S. Hardy, W. Henecka, et al., Entity resolution and federated learning get a federated resolution, *ArXiv Preprint ArXiv:1803.04035*, March 2018. http://arxiv.org/abs/1803.04035

S. J. Pan and Q. Yang, A survey on transfer learning, *IEEE Transactions on Knowledge and Data Engineering*, 22(10):1345–1359, 2010. DOI: 10.1109/tkde.2009.191. 9, 83, 84

S. J. Pan, I. W. Tsang, J. T. Kwok, and Q. Yang, Domain adaptation via transfer component analysis, *Proc. of the 21st International Joint Conference on Artificial Intelligence*, pp. 1187–1192, 2009. DOI: 10.1109/tnn.2010.2091281. 84

J. Augustine, N. Chen, E. Elkind, et al., Dynamics of profit-sharing games, *Internet Mathematics*, 1:1–22, 2015. DOI: 10.1080/15427951.2013.830164. 97

S. Barbara and M. Jackson, Maximin, leximin, and the protective criterion: Characterizations and comparisons, *Journal of Economic Theory*, 46(1):34–44, 1988. DOI: 10.1016/0022-0531(88)90148-2.

The Belmont report. *Technical Report*, National commission for the protection of human subjects of biomedical and behavioral research, *Department of Health, Education and Welfare*, United States Government Printing Office, Washington, DC, 1978. 101

G. Christodoulou, K. Mehlhorn, and E. Pyrga, Improving the price of anarchy for selfish routing via coordination mechanisms, In *ESA*, pp. 119–130, 2011. DOI: 10.1007/978-3-642-23719-5_11.

Regulation (EU) 2016/679 of the European Parliament and of the Council 27 April 2016 on the protection of natural persons with regard to the processing of personal data and on the free movement of such data, and repealing directive 95/46/ec (general data protection regulation), *Technical Report*, European Union, 2016.

B. Faltings, G. Radanovic, and R. Brachman, *Game Theory for Data Science: Eliciting Truthful Information*, Morgan & Claypool Publishers, 2017. DOI: 10.2200/s00788ed1v01y201707aim035.

S. Gollapudi, K. Kollias, D. Panigrahi, et al., Profit sharing and efficiency in utility games, In *ESA*, pp. 1–16, 2017. 96, 97

K. Kollias and T. Roughgarden, Restoring pure equilibria to weighted congestion games, *ACM Transactions on Economics and Computation*, 3(4):21:1–21:24, 2015. DOI: 10.1145/2781678.

T. Luo, S. S. Kanhere, J. Huang, et al., Sustainable incentives for mobile crowdsensing: Auctions, lotteries, and trust and reputation systems, *IEEE Communications Magazine*, 55(3):68–74, 2017. DOI: 10.1109/mcom.2017.1600746cm.

J. R. Marden and A. Wierman, Distributed welfare games, *Operations Research*, 61(1):155–168, 2013. DOI: 10.1287/opre.1120.1137.

M. J. Neely, *Stochastic Network Optimization with Application to Communication and Queueing Systems*, Morgan & Claypool Publishers, 2010. DOI: 10.2200/s00271ed1v01y201006cnt007. 101

R. Shokri and V. Shmatikov, Privacy-preserving deep learning, In *Proc. of the ACM SIGSAC Conference on Computer and Communications Security (CCS)*, pp. 1310–1321, October 2015. DOI: 10.1109/allerton.2015.7447103. 43, 44, 47, 49, 89

P. von Falkenhausen and T. Harks, Optimal cost sharing protocols for scheduling games, In *Proc. of the 12th ACM Conference on Electronic Commerce (EC)*, pp. 285–294, June 2011. DOI: 10.1145/1993574.1993618.

S. Yang, F. Wu, S. Tang, et al., On designing data quality-aware truth estimation and surplus sharing method for mobile crowdsensing, *IEEE Journal on Selected Areas in Communications*, 35(4):832–847, 2017. DOI: 10.1109/jsac.2017.2676898. 96

H. Yu, C. Miao, Z. Shen, et al., Efficient task sub-delegation for crowdsourcing, In *29th AAAI Conference on Artificial Intelligence*, pp. 1305–1311, February 2015. 101

H. Yu, C. Miao, C. Leung, et al., Mitigating herding in hierarchical crowdsourcing networks, *Scientific Reports*, 6(4), 2016. DOI: 10.1038/s41598-016-0011-6. 101

H. Yu, Z. Shen, C. Miao, et al., Building ethics into artificial intelligence, *ArXiv Preprint ArXiv:1812.02953*, December 2018. http://arxiv.org/abs/1812.02953 101

H. Yu, C. Miao, Y. Zheng, et al., Ethically aligned opportunistic scheduling for productive laziness, *ArXiv Preprint ArXiv:1901.00298*, January 2019. http://arxiv.org/abs/1901.00298 DOI: 10.1145/3306618.3314240. 101

S. Ruder, *Neural Transfer Learning for Natural Language Processing*, National University of Ireland, Galway, 2019. 92

E. Bagdasaryan, A. Veit, Y. Hua, et al., ImageNet: A large-scale hierarchical image database, In *IEEE Conference on Computer Vision and Pattern Recognition*, pp. 248–255, 2009. DOI: 10.1109/CVPR.2009.5206848. 92

H. B. McMahan and D. Ramage, Federated learning: Collaborative machine learning without centralized training data, April 2017. https://ai.googleblog.com/2017/04/federated-learning-collaborative.html 69, 139

K. Xu, W. Hu, J. Leskovec, et al., How powerful are graph neural networks?, *ArXiv Preprint ArXiv:1810.00826*, October 2018. http://arxiv.org/abs/1810.00826

D. Preuveneers, V. Rimmer, I. Tsingenopoulos, et al., Chained anomaly detection models for federated learning: An intrusion detection case study, In *Applied Sciences*, December 2018. DOI: 10.3390/app8122663. 21, 65, 140

Y. Zheng, F. Liu, and H. Hsieh, U-Air: When urban air quality inference meets big data, In *Proc. of the 19th ACM SIGKDD International Conference on Knowledge Discovery and Data Mining (KDD)*, pp. 1436–1444, New York, 2013. https://doi.org/10.1145/2487575.2488188 DOI: 10.1145/2487575.2488188. 136

CNNIC publishes the 41st statistical report on China's Internet development in China. https://www.lexology.com/library/detail.aspx?g=911ae57f-50da-4c53-ab75-2376272b2021 139

eMarketer publishes Worldwide Internet and mobile users: eMarketer's updated estimates and forecast for 2017–2021. https://www.emarketer.com/Report/Worldwide-Internet-Mobile-Users-eMarketers-Updated-Estimates-Forecast-20172021/2002147 139

R. S. Sutton and A. G. Barto, *Introduction to Reinforcement Learning*, MIT Press, 1998. 121

G. A. Rummery and M. Niranjan, *On-Line Q-Learning Using Connectionist Systems*, Cambridge University Engineering Department, 1994. 124

C. Watkins and P. Dayan, Q-learning, In *Machine Learning*, pp. 279–292, 1992. DOI: 10.1007/bf00992698. 124

J. Chen, X. Pan, R. Monga, et al., Revisiting distributed synchronous SGD, March 2017. http://arxiv.org/abs/1604.00981 33, 51, 126

V. Mnih, A. P. Badia, M. Mirza, et al., Asynchronous methods for deep reinforcement learning, In *Proc. of the 33rd International Conference on Machine Learning*, pp. 1928–1937, June 2016. 125, 126

A. Nair, P. Srinivasan, S. Blackwell, et al., Massively parallel methods for deep reinforcement learning, July 2015. http://arxiv.org/abs/1507.04296 125

A. V. Clemente, H. N. Castejón, and A. Chandra, Efficient parallel methods for deep reinforcement learning, May 2017. http://arxiv.org/abs/1705.04862 126

V. Chen, V. Pastro, and M. Raykova, Secure computation for machine learning with SPDZ, January 2019. https://arxiv.org/abs/1901.00329 26

R. Cramer, I. Damgård, D. Escudero, et al., SPDZ$_{2k}$: Efficient MPC mod 2k for dishonest majority, In *Annual International Cryptology Conference*, pp. 769–798, Springer, 2018. 26

R. Gilad-Bachrach, N. Dowlin, K. Laine, et al., CryptoNets: Applying neural networks to encrypted data with high throughput and accuracy, In *International Conference on Machine Learning*, pp. 201–210, 2016. 28

C. Juvekar, V. Vaikuntanathan, and A. Chandrakasan, Gazelle: A low latency framework for secure neural network inference, In *USENIX Security Symposium*, 2018. 29

D. Chai, L. Wang, K. Chen, and Q. Yang, Secure federated matrix factorization, June 2019. https://arxiv.org/abs/1906.05108 29

N. Phan, X. Wu, and D. Dou, Preserving differential privacy in conutional deep belief networks, In *Machine Learning*, 106(9):1681–1704, October 2017. DOI: 10.1007/s10994-017-5656-2. 32

A. Triastcyn and B. Faltings, Generating differentially private datasets using GANs, February 2018. https://openreview.net/pdf?id=rJv4XWZA- 32

L. Yu, L. Liu, C. Pu, et al., Differentially private model publishing for deep learning, May 2019. https://arxiv.org/abs/1904.02200 DOI: 10.1109/sp.2019.00019. 32

X. Chen, T. Chen, H. Sun, et al., Distributed training with heterogeneous data: Bridging median- and mean-based algorithms, June 2019. https://arxiv.org/abs/1906.01736 56, 63

L. Li, W. Xu, T. Chen, et al., RSA: Byzantine-robust stochastic aggregation methods for distributed learning from heterogeneous datasets, November 2018. https://arxiv.org/abs/1811.03761 DOI: 10.1609/aaai.v33i01.33011544. 56

M. Duan, Astraea: Self-balancing federated learning for improving classification accuracy of mobile deep learning applications, July 2019. https://arxiv.org/abs/1907.01132 56

iResearch, Report on China's smart cities development, 2019. https://www.iresearch.com.cn/Detail/report?id=3350&isfree=0 137

A. Chen, IBM's Watson gave unsafe recommendations for treating cancer, July 2018. `https://www.theverge.com/2018/7/26/17619382/ibms-watson-cancer-ai-healthcare-science` 134

L. Mearian, Did IBM overhype Watson health's AI promise?, November 2018. `https://www.computerworld.com/article/3321138/did-ibm-put-too-much-stock-in-watson-health-too-soon.html` 134

A. van den Oord, S.Dieleman, H. Zen, et al., WaveNet: A generative model for raw audio, September 2016. `https://arxiv.org/abs/1609.03499` 112

F. Baldimtsi, D. Papadopoulos, S. Papadopoulos, et al., Server-aided secure computation with off-line parties, In *Computer Security—ESORICS*, pp. 103–123, 2017. DOI: 10.1007/978-3-319-66402-6_8. 81

R. Bost, R. A. Popa, S. Tu, and S. Goldwasser, Machine learning classification over encrypted data, In *NDSS*, pp. 103–123, February 2015. DOI: 10.14722/ndss.2015.23241. 81

Covington and Burling LLP, Inside privacy: Updates on developments in data privacy and cybersecurity, July 2019. `https://www.insideprivacy.com/uncategorized/china-releases-draft-measures-for-the-administration-of-data-security/` 154

H. Guo, R. Tang, Y. Ye, et al., DeepFM: A factorization-machine based neural network for CTR prediction, In *Proc. of the 26th International Joint Conference on Artificial Intelligence, IJCAI*, pp. 1725–1731, August 2017. `https://doi.org/10.24963/ijcai.2017/239` DOI: 10.24963/ijcai.2017/239. 118

O. Habachi, M.-A. Adjif, and J.-P. Cances, Fast uplink grant for NOMA: A federated learning based approach, March 2019. `https://arxiv.org/abs/1904.07975` 141

S. Niknam, H. S. Dhillon, and J. H. Reed, Federated learning for wireless communications: Motivation, opportunities and challenges, September 2019. `https://arxiv.org/abs/1908.06847` 141

K. B. Letaief, W. Chen, Y. Shi, et al., The roadmap to 6G—AI empowered wireless networks, July 2019. `https://arxiv.org/abs/1904.11686` DOI: 10.1109/mcom.2019.1900271. 141

Z. Zhou, X. Chen, E. Li, et al., Edge intelligence: Paving the last mile of artificial intelligence with edge computing, May 2019. `https://arxiv.org/abs/1905.10083` DOI: 10.1109/jproc.2019.2918951. 141

G. Zhu, D. Liu, Y. Du, et al., Towards an intelligent edge: Wireless communication meets machine learning, September 2018. `https://arxiv.org/abs/1809.00343` 141

S. Samarakoon, M. Bennis, W. Saad, and M. Debbah, Federated learning for ultra-reliable low-latency V2V communication, In *Proc. of the IEEE Globecom*, 2018. DOI: 10.1109/glocom.2018.8647927. 141

E. Jeong, S. Oh, H. Kim, et al., Communication-efficient on-device machine learning: Federated distillation and augmentation under non-IID private data, *NIPS Workshop*, Montreal, Canada, 2018. 141

M. Bennis, Trends and challenges of federated learning in the 5G network, July 2019. https://www.comsoc.org/publications/ctn/edging-towards-smarter-network-opportunities-and-challenges-federated-learning 141

J. Park, S. Samarakoon, M. Bennis, and M. Debbah, Wireless network intelligence at the edge, September 2019. https://arxiv.org/abs/1812.02858 DOI: 10.1109/jproc.2019.2941458. 141

Q. Li, Z. Wen, and B. He, Federated learning systems: Vision, hype and reality for data privacy and protection, July 2019a. http://arxiv.org/abs/1907.09693 66

T. Li, A. K. Sahu, A. Talwalkar, and V. Smith, Federated learning: Challenges, methods, and future directions, August 2019. https://arxiv.org/abs/1908.07873 66

F. Mo and H. Haddadi, Efficient and private federated learning using TEE, March 2019. https://eurosys2019.org/wp-content/uploads/2019/03/eurosys19posters-abstract66.pdf 21

R. C. Geyer, T. Klein, and M. Nabi, Differentially private federated learning: A client level perspective, March 2018. https://arxiv.org/abs/1712.07557 32

M. Al-Rubaie and J. M. Chang, Reconstruction attacks against mobile-based continuous authentication systems in the cloud, In *IEEE Transactions on Information Forensics and Security*, 11(12):2648–2663, 2016. DOI: 10.1109/tifs.2016.2594132. 20

M. Fredrikson, S. Jha, and T. Ristenpart, Model inversion attacks that exploit confidence information and basic countermeasures, In *Proc. of the 22nd ACM SIGSAC Conference on Computer and Communications Security*, pages 1322–1333, 2015. DOI: 10.1145/2810103.2813677. 19

P. Xie, B. Wu, and G. Sun, BAYHENN: Combining Bayesian deep learning and homomorphic encryption for secure DNN inference, In *Proc. of the 28th International Joint Conference on Artificial Intelligence, IJCAI*, pages 4831–4837, Macao, China, August 10–16, 2019. DOI: 10.24963/ijcai.2019/671. 20

R. Shokri, M. Stronati, C. Song, and V. Shmatikov, Membership inference attacks against machine learning models, In *IEEE Symposium on Security and Privacy (SP)*, pages 3–18, 2017. DOI: 10.1109/sp.2017.41. 20

D. Wang, L. Zhang, N. Ma, and Xiaobo Li, Two secret sharing schemes based on Boolean operations, *Pattern Recognition*, 40(10):2776–2, 2017. DOI: 10.1016/j.patcog.2006.11.018. 24

K. Xu, H. Mi, D. Feng, et al., Collaborative deep learning across multiple data centers, October 2018. https://arxiv.org/abs/1810.06877 52, 55

I. Cano, M. Weimer, D. Mahajan, et al., Towards geo-distributed machine learning, March 2016. https://arxiv.org/abs/1603.09035 52, 55

A. Reisizadeh, A. Mokhtari, H. Hassani, et al., FedPAQ: A Communication-efficient federated learning method with periodic averaging and quantization, October 2019. https://arxiv.org/abs/1909.13014 63

L. Wang, W. Wang, and B. Li, CMFL: Mitigating communication overhead for federated learning, In *Proc. of the 39th IEEE International Conference on Distributed Computing Systems (ICDCS)*, July 2019. 64

K. Hsieh, A. Harlap, N. Vijaykumar, et al., Gaia: Geo-distributed machine learning approaching LAN speeds, In *NSDI*, pp. 629–647, 2017. 52, 64

A. Zhang, Z. C. Lipton, M. Li, and A. J. Smola, Dive into deep learning, October 2019. https://en.d2l.ai/d2l-en.pdf 57

T.-Y. Liu, W. Chen, T. Wang, and F. Gao, *Distributed Machine Learning: Theories, Algorithms, and Systems*, China Machine Press, September 2018. (In Chinese, ISBN 978-7-111-60918-6.) 33

J. Redmon, S. Divvala, R. Girshick, and A. Farhadi, You only look once: Unified, real-time object detection, May 2016. Available: https://arxiv.org/abs/1506.02640 108

H. Yu, Z. Liu, Y. Liu, T. Chen, M. Cong, X. Weng, D. Niyato, and Q. Yang, A fairness-aware incentive scheme for federated earning, In *Proc. of the 3rd AAAI/ACM Conference on Artificial Intelligence, Ethics, and Society (AIES-20)*, 2020. 95

F. Haddadpour, M. M. Kamani, M. Mahdavi, and V. R. Cadambe, Local SGD with periodic averaging: Tighter analysis and adaptive synchronization, October 2019. https://arxiv.org/abs/1910.13598 55

P. Kairouz, H.B. McMahan, B. Avent, et al., Advances and open problems in federated learning, December 2019. https://arxiv.org/abs/1912.04977 7, 49

Authors' Biographies

QIANG YANG

Qiang Yang is the head of the AI department at WeBank (Chief AI Officer) and Chair Professor at the Computer Science and Engineering (CSE) Department of the Hong Kong University of Science and Technology (HKUST), where he was a former head of CSE Department and founding director of the Big Data Institute (2015–2018). His research interests include AI, machine learning, and data mining, especially in transfer learning, automated planning, federated learning, and case-based reasoning. He is a fellow of several international societies, including ACM, AAAI, IEEE, IAPR, and AAAS. He received his Ph.D. in Computer Science in 1989 and his M.Sc. in Astrophysics in 1985, both from the University of Maryland, College Park. He obtained his B.Sc. in Astrophysics from Peking University in 1982. He had been a faculty member at the University of Waterloo (1989–1995) and Simon Fraser University (1995–2001). He was the founding Editor-in-Chief of the *ACM Transactions on Intelligent Systems and Technology (ACM TIST)* and *IEEE Transactions on Big Data (IEEE TBD)*. He served as the President of International Joint Conference on AI (IJCAI, 2017–2019) and an executive council member of Association for the Advancement of AI (AAAI, 2016–2020). Qiang Yang is a recipient of several awards, including the 2004/2005 ACM KDDCUP Championship, the ACM SIGKDD Distinguished Service Award (2017), and AAAI Innovative AI Applications Award (2016). He was the founding director of Huawei's Noah's Ark Lab (2012–2014) and a co-founder of 4Paradigm Corp, an AI platform company. He is an author of several books including *Intelligent Planning* (Springer), *Crafting Your Research Future* (Morgan & Claypool), and *Constraint-based Design Recovery for Software Engineering* (Springer).

YANG LIU

Yang Liu is a Senior Researcher in the AI Department of WeBank, China. Her research interests include machine learning, federated learning, transfer learning, multi-agent systems, statistical mechanics, and applications of these technologies in the financial industry. She received her Ph.D. from Princeton University in 2012 and her Bachelor's degree from Tsinghua University in 2007. She holds multiple patents. Her research has been published in leading scientific journals such as *ACM TIST* and *Nature*.

YONG CHENG

Yong Cheng is currently a Senior Researcher in the AI Department of WeBank, Shenzhen, China. Previously, he had worked in Huawei Technologies Co., Ltd. (Shenzhen) as a Senior Engineer, and in Bell Labs Germany as a Senior Researcher. Yong had also worked as a Researcher in the Huawei-HKUST Innovation Laboratory, Hong Kong. His research interests and expertise mainly include Deep Learning, Federated Learning, Computer Vision and OCR, Mathematical Optimization and Algorithms, Distributed Computing, as well as Mixed-Integer Programming. He has published more than 20 journal and conference papers and filed more than 40 patents. Yong received the B.Eng. (1st class honors), MPhil, and Ph.D. (1st class honors) degrees from Zhejiang University (ZJU), Hangzhou, PR China, the Hong Kong University of Science and Technology (HKUST), Hong Kong, and Technische Universität Darmstadt (TU Darmstadt), Darmstadt, Germany, in 2006, 2010, and 2013, respectively. He received the best Ph.D. thesis award of TU Darmstadt in 2014, and the best B.Eng. thesis award of ZJU in 2006. Yong gave a tutorial on "Mixed-Integer Conic Programming" at ICASSP'15, and he was the PC Member of FML'19 (in conjunction with IJCAI'19).

YAN KANG

Yan Kang is a Senior Researcher in the AI department of Webank in Shenzhen, China. His work is focusing on the research and implementation of privacy-preserving machine learning and federated transfer learning techniques. He received M.S. and Ph.D. degrees in Computer Science from the University of Maryland, Baltimore County, USA. His Ph.D. work was awarded a doctoral fellowship and centered around machine learning and semantic web for heterogeneous data integration. During his graduate work, he participated in multiple projects collaborating with the National Institute of Standards and Technology (NIST) and the National Science Foundation (NSF) for designing and developing ontology integration systems. He also has adequate experiences in commercial software projects. Before joining WeBank, he had been working for Stardog Union Inc. and Cerner Corporation for more than four years on system design and implementation. His github page is `https://github.com/yankang18`.

TIANJIAN CHEN

Tianjian Chen is the Deputy General Manager of the AI Department of WeBank, China. He is now responsible for building the Banking Intelligence Ecosystem based on Federated Learning Technology. Before joining WeBank, he was the Chief Architect of Baidu Finance, Principal Architect of Baidu. Tianjian has over 12 years of experience in large-scale distributed system design and enabling technology innovations in various application fields, such as web search engine, peer-to-peer storage, genomics, recommender system, digital banking, and machine learning.

HAN YU

Han Yu is a Nanyang Assistant Professor (NAP) in the School of Computer Science and Engineering (SCSE), Nanyang Technological University (NTU), Singapore. Between 2015 and 2018, he held the prestigious Lee Kuan Yew Post-Doctoral Fellowship (LKY PDF). Before joining NTU, he worked as an Embedded Software Engineer at Hewlett-Packard (HP) Pte Ltd, Singapore. He obtained his Ph.D. in Computer Science from NTU in 2014. His research focuses on online convex optimization, ethical AI, federated learning, and their applications in complex collaborative systems such as crowdsourcing. He has published over 120 research papers leading international conferences and journals and won multiple research awards (`http://hanyu.sg`).

Printed in the United States
by Baker & Taylor Publisher Services